Morgellon's Syndrome

Evidence of a Microorganism Causing an Unexplained Dermopathy

Morgellon's Syndrome

Evidence of a Microorganism Causing an Unexplained Dermopathy

Dr. David Conroy MD, LLM
Lansdowne Medical Clinic
Toronto
Canada

CRC Press
Taylor & Francis Group
Boca Raton London New York

CRC Press is an imprint of the
Taylor & Francis Group, an **informa** business

CRC Press
Taylor & Francis Group
6000 Broken Sound Parkway NW, Suite 300
Boca Raton, FL 33487-2742

First issued in hardback 2017

ISBN-13: 978-1-57808-673-3 (pbk)
ISBN-13: 978-1-138-41676-5 (hbk)

Library of Congress Cataloging-in-Publication Data

Conroy, David, 1964-
Morgellon's syndrome : evidence of a microorganism causing an unexplained dermopathy / David Conroy.
p.; cm.
Includes bibliographical references and index.
ISBN 978-1-57808-673-3 (pbk. : alk. paper)
1. Neurocutaneous disorders--Case studies. 2. Dermatomycoses--Case studies. I. Title.
[DNLM: 1. Ectoparasitic Infestations--diagnosis--Case Reports. 2. Morgellons Disease--etiology--Case Reports. 3. Cocaine-Related Disorders--complications--Case Reports. 4. Delusions--diagnosis--Case Reports. 5. Parasites--Case Reports. WC 900 C754m 2009]
RL751.C66 2009
616.5'79--dc22

2009040660

Visit the Taylor & Francis Web site at
http://www.taylorandfrancis.com

and the CRC Press Web site at
http://www.crcpress.com

Preface

There are thousands of people diagnosed with delusional parasitosis by physicians. The treatment involves an antipsychotic medication with many serious adverse effects some of them potentially fatal. Many of the people afflicted with this disease reject the diagnosis believing that their symptoms are caused by an as of yet unidentified microorganism.

The book proposes that perhaps there is more to the pathophysiology of persons diagnosed with delusional parasitosis than delusions brought on by mental illness and stimulant drug use. It shows several phenomena that constitute evidence that there is microorganism that may be responsible for many cases of unexplained dermopathies.

The book is accompanied by a DVD containing all of the photographs and videos referred to in the book. The reader is encouraged to download the DVD and use photograph and video software to magnify the photographs and to perform their own frame by frame analysis of the videos. I used the Sony Picture Utility software provided with my camera.

It is hoped that some more definitive research is performed using the information presented in my book. I performed my investigation on my own dime with nothing but a camera, a microscope and dogged determination. The importance of investigating the cause of unexplained dermopathies by a well-funded institution is clear - the diagnosis of having a mental illness is one of the hardest diagnoses to live with and especially unjust if it is wrong.

David Conroy

Supplementary Resources Disclaimer

Additional resources were previously made available for this title on DVD. However, as DVD has become a less accessible format, all resources have been moved to a more convenient online download option.

You can find these resources available here: www.routledge.com/9781138416765

Please note: Where this title mentions the associated disc, please use the downloadable resources instead.

Contents

Introduction

Every so often, scientific knowledge undergoes a revision. What was once thought to be true is cast into doubt and sometimes new knowledge replaces the old. This may be the eventual fate of delusional parasitosis.

Prior to 1994, peptic ulcers were thought to be caused by the hypersecretion of gastric acid caused by increased but normal physiologic processes. The accepted causes at the time of the hypersecretion of gastric acid included caffeine, spices, alcohol, cigarette smoke and psychological stress. In 1982, Australian physicians Robin Warren and Barry Marshall cultured Helicobacter pylori on a forgotten Petri dish. To finally silence skeptics, Marshall drank a beaker of Helicobacter pylori to demonstrate the role of this bacteria in the pathophysiology of gastritis and peptic ulcers. Then in 1994, the National Institutes of Health (USA) recognized that most cases of gastric ulcers were caused by *Helicobacter pylori.* Warren and Marshall were awarded the Nobel Prize in Medicine in 2005 for their efforts.[1]

The high prevalence of Kaposi's sarcoma, a form of dermatological cancer, in AIDS patients pointed to a viral etiology, prompting research into a possible causative agent. Then in 1994, Kaposi's sarcoma was determined to be caused by Human herpesvirus 8 (now known as Kaposi's sarcoma-associated herpesvirus or KSHV) by a team of Columbia University researchers.[2]

It is now time to reconsider the etiology of some or possibly all the cases of delusional parasitosis. After a 15 month investigation, evidence of a microorganism being the causative agent may have been discovered. Whether or not this microorganism turns out to be the etiology of all cases of this disease, it is most likely the cause of some of the cases, particularly in those cases involving cocaine users.

Delusional parasitosis is defined as "a form of psychosis of false belief, a "loss of contact with reality". In delusional parasitosis, sufferers have a strong delusional belief that they are infested with parasites, whereas in reality no such parasites are present."[3] These patients present to dermatologists with specimens consisting of

---

[1]http://en.wikipedia.org/wiki/Helicobacter_pylori
[2]http://en.wikipedia.org/wiki/Kaposi's_sarcoma
[3]http://en.wikipedia.org/wiki/Delusional_parasitosis

lint, skin or other debris contained in plastic saran wrap, celluloid tape, matchboxes or other containers (the so-called "matchbox sign").

Persons diagnosed with delusional parasitosis present with the belief that parasites are crawling under their skin and sometimes exiting the dermis. Some claim that materials such as hair and lint exit from their epidermis. The disease is thought to be the result of psychotic delusions precipitated by mental illness or stimulant drug use. In keeping with their perceived delusional origin, these symptoms are currently treated with the antipsychotic medications risperidone and olanzapine.[4]

The research reported in this book proves that inert objects such as hair, feathers and lint and non-human white particles are entering the epidermis. Hairs, a feather and white particles are also seen exiting the dermis and by extension, lint is most likely also exiting the epidermis. The theories of how and why are discussed in the analysis found in the final chapters of the book.

The investigation undertaken here examines the possibility that these symptoms are not the result of delusions precipitated by the stimulant cocaine. Although the case study presented here is a cocaine user, this may be incidental. It is unknown whether cocaine is a prerequisite to be infected by the microorganism discussed here and other persons diagnosed with delusional parasitosis have a separate etiology, or whether cocaine only makes the host more susceptible than other sufferers of delusional parasitosis to the same microorganism infecting non-users of cocaine. Other possibilities include the possibility that cocaine users are more aware of their symptoms or that the organism is more active in the presence of cocaine and therefore more apparent and visible.

After global warming, cocaine is arguably the most deleterious substance on the planet. In recent years, ten thousand Mexicans have died in violence attributed to the drug trade. Colombians have been suffering from the terror accompanying drug trafficking for decades. The human capacities of millions of Europeans, Americans and Canadians have been diminished or totally eliminated by their addiction to cocaine. It goes without saying that the world would be a much more humane place if this scourge could be reduced or even eliminated.

I see the dehumanizing affects of cocaine addiction every day in my clinic located in Parkdale, a drug affected neighborhood in Toronto. I see abused and neglected children, usually brought in by social workers and foster parents, whose parents are lost to their addictions. I am deeply saddened by formerly productive human beings spending their whole lives in pursuit of the temporary euphoria provided by cocaine. In the remote chance that an investigation into a possible microorganism infecting cocaine users may help reduce cocaine use, I decided to

[4]Noah S Scheinfeld, "Delusions of Parasitosis." Available at http://emedicine.medscape.com/article/1121818.

take the opportunity to investigate the strange complaints made by someone who I suspected was consuming cocaine. Further inquiries confirmed that the patient was in fact consuming cocaine.

With that background information, it was decided to investigate the complaints made by a confirmed user of cocaine. The complaints consisted of the paradoxical appearances of small patches of paroxysmal epidermal pruritis, the paroxysmal paradoxical appearance of hair inside of socks and the sudden appearance of abnormal domesticated feline behavior. These complaints were investigated in both the human subject and his cat over a 16 month period using a video camera and microscope.

Several totally unique phenomena were recorded including hair that appears to be infected with what might turn out to be a fungal-type of microscopic organism. The microorganism appears to have its own propulsion system and the ability to enter and exit the epidermis, cornea and sclera rapidly. These specific phenomena are present only in cocaine users and are elicited only in the presence of cocaine, either after consumption by the host or as fumes in the surrounding atmosphere.

The distinct possibility exists that the subject has Morgellon's disease[5] which is somehow amplified by the use of cocaine. Morgellon's disease is the name given to many cases of delusional parasitosis by people that believe that many if not all cases of delusional parasitosis are not caused by delusions but have an infectious etiology. The Centers for Disease Control and Prevention (CDC) began an investigation into this unexplainable dermopathy. At the time of writing this book, phase one of the study was completed and phase two of the investigation is underway. Phase one consisted of subject evaluation and selection while phase two consists of sample retrieval and analysis.

Important questions remain. How widespread is this organism in the population? Is it responsible for, or contribute to, known or unknown diseases such as Morgellon's? Are there any medicinal or industrial uses for this organism? And most importantly, can this organism consume cocaine and more specifically, can it be utilized to eradicate cocaine plants? Further research is needed in order to answer these and other questions that may arise.

[5]"Mysterious 'Morgellon's' disease prompts US investigation," Emma Marris, Nature Medicine 12, 982 (1 September 2006).

1

Why the Investigation was Started

This investigation was undertaken after several complaints of dermatological abnormalities were made by cocaine users. The first complaint category was a pruritic rash that was more pronounced while ingesting cocaine. The rash was most pronounced on the extremities especially the distal aspects of the lower limbs.

The second category of complaints was that they believed that mites were shooting off the ceiling, landing on their heads and then burrowing into their scalps. The complainant did not have to be actively ingesting cocaine to experience this phenomenon, however, the complainant had to be in an area where cocaine was chronically present. Non-cocaine users residing in the same household did not experience the same symptoms. One complainant reported that her partner had taken to sleeping in a tent set up in the living room in an effort to avoid this symptom. Another complainant reported that she could hear the mites hitting the floor as she walked by and that her cat was hiding under furniture to avoid them.

A third type of complaint was that mites were shooting off their cocaine pipes while the cocaine was being vaporized and then burrowing into their faces. The same phenomenon was described by most cocaine users not as burrowing mites, but rather as impurities in bad cocaine. These complainants would occasionally have several black spots on their faces after using the "bad" cocaine. This complaint only occurred while using bad, crumbly cocaine bicarbonate and was not present when using good quality, presumably uninfected, cocaine.

The last symptom shared by many complainants was that filamentous things were coming out of their scalps while actively engaged in smoking cocaine. While ingesting cocaine or shortly thereafter, the complainants reported that a small patch of their scalps would begin to tingle. On occasion, within a couple of minutes a large piece of lint would fall out of their hair. Similar symptoms also occurred on their backs but without the expulsion of lint. Instead it felt like a large hair was exiting their backs, sometimes lifting their shirts off their backs.

It should be stated here that two of the complainants were not cocaine users. One was a grandmother concerned about a household infestation. The other was a middle-aged man experiencing the sensation of formication and white and black materials exiting his skin. These would be non-stimulant using cases of unexplained dermopathies, possibly Morgellon's.

A telephone consultation with an infectious disease specialist, one of the world's leading experts on the controversial "cocaine mite," was made in August 2007. He reported that the dermatological symptoms experienced by cocaine smokers were actually delusions brought on by the cerebral effects of cocaine which included hallucinations and paranoia. He went on to say that the specimens he received from the hair of his patients were, in reality, pieces of lint. The specialist reported that the diagnosis was delusional parasitosis and the treatment was the administration of the anti-psychotic and major tranquilizer medication risperidone at a dose of 1 mg by mouth daily. He went on to say that after the risperidone treatment was initiated, the patients reported a reduction or cessation of their symptoms.

The effectiveness of risperidone in reducing the reported dermatological symptoms by cocaine users may appear to confirm the etiology of these symptoms as psychotic hallucinations. There are several alternative explanations for the reduction in the reported symptomatology that occurs with risperidone therapy.

The first possible alternative reason is that risperidone is also known as a major tranquilizer and the reduction in symptoms may be due to the tranquilizing effects of the drug. The Compendium of Pharmaceuticals and Specialties lists somnolence as a common side effect of risperidone.

Secondly, the symptoms may become less bothersome with time without any treatment. The subject of this investigation reported a reduction in anxiety associated with the symptomatology as time progressed. It appears that the mechanism for the reduction in symptoms is similar to the mechanism of action seen in desensitization therapy for anxiety disorders. The initial anxiety that prompts the cocaine addict to seek medical attention seems to be rooted in the belief that they are infested with an internal parasite that may be causing serious damage to their health. After several months of experiencing the sensation of focal paroxysmal paradoxical pruritis without adverse consequences, the research subject stated that the anxiety and concern over the tingling subsided.

Thirdly, the effects of risperidone on symptom reduction may be psychological. The majority of cocaine users consulted for this investigation believed that indeed these symptoms were a result of the psychotic effects of cocaine. The reason for this belief was most likely that consultations with infectious disease specialists and dermatologists resulted in diagnoses of delusional parasitosis in those few who sought treatment. Those cocaine addicts with the diagnosis then relate the information to their colleagues and the theory of the symptoms arising from hallucinations is widely accepted partly because they acknowledge that cocaine causes other delusions as well as illusions and paranoia. The belief that the sensation of parasitosis was a delusion caused by the cerebral effects of cocaine may be a sort of reassurance to the infected individual.

At this point, a decision had to be made whether or not to investigate whether the "cocaine mite" was in reality a result of psychotic hallucinatory delusions brought on by the ingestion of cocaine or the result of an undiscovered parasite. This was a very difficult decision to make. Cocaine is illegal and great risks were involved in researching this substance without approval. In addition, the decision to supply an individual an extremely dangerous and addictive substance was also of questionable ethics.

The decision to commence an illegal research project was made for essentially this reason: cocaine is responsible for the destruction of millions of lives worldwide. The substance does not only affect the individual user but the whole of society. Child abuse and neglect, murdered police officers and innocent bystanders, and the loss of the user's own humanity were all compelling reasons to investigate these complaints. If this research and the consequent possibility of me becoming publicly disgraced and possibly even incarcerated could save even one life it was worth the effort. The cocaine trade is probably the root cause of most of the crime and gang violence in the developed world. Colombia and Mexico have been devastated by the violence associated with the production and distribution of the drug. It was decided that the possibility of discovering an organism that could be the beginning of the end for cocaine was important enough to commence an investigation despite all the inherent risks.

With the ethical dilemma solved, or more accurately, rationalized, an anonymous research subject was selected and supplied with between 600mg and 1000mg of cocaine bicarbonate daily and the results were videotaped and specimens collected and examined with a microscope.

The initial observation indicating that a newly discovered form of life exists was seen in the research subject's domestic feline. For no apparent reason, a feline of the tuxedo type developed several previously unnoticed behaviors. First, the subject feline began avoiding open areas of the habitation preferring to ambulate throughout the house while remaining beneath furniture. Secondly, the subject cat began exhibiting paroxysms of scratching and gnawing various parts of his body. The third new behavior was that the cat developed paroxysmal tearing.

Subsequently, the human research subject noticed hair and white matter emerging from his skin. Several specimens were sent to parasitology and no organisms were detected. High definition videos of the subject reveal several abnormal phenomena which most likely are caused by a unique previously undiscovered microscopic organism.

2

Materials and Methodology

To test the hypothesis that cocaine was responsible for this infection, cocaine was incinerated on a metal copper pipe and the vaporized cocaine inhaled and then blown onto the feet. The decision to expose the feet to the cocaine vapors was made for two reasons. First of all, the tingling in the hair as well as the expulsion of materials from the scalp occurred only while the subject was actively vaporizing cocaine. The presence of cocaine seemed to stimulate whatever was causing these phenomena. It was logical then to determine whether high concentrations of cocaine molecules in the surrounding environment had any effect.

Secondly, the decision to expose the feet rather than, say, the arm was made because the feet were the sites where hair was thought to be spontaneously appearing inside of the socks. Initially, the hope was that the high concentrations of cocaine vapors would induce the appearance of hair emerging from the soles of the feet. However, despite expectations based on the above discussion, the spontaneous emergence of hair from the feet was not witnessed throughout the duration of the investigation.

The statement that no human hair was seen emerging from the feet during the entire investigation is not entirely accurate. First of all, a strand of hair was seen protruding from the plantar surface of the foot which then appeared to elongate (see chapter 9). Secondly, hair was videotaped entering and then subsequently exiting the epidermis (see chapter 4). Arguably, these two types of observations could be classified as hair coming out of the epidermis of the feet. It is probably better to say that no hair was observed exiting the epidermis of the feet prior to the hair first being observed entering the epidermis of the feet and no hair was observed exiting the feet prior to their appearance on the feet. This could be explained by assuming that the hair only exits the feet while the feet are in complete darkness. While no hair was seen exiting the feet, per se, lots of hair was observed entering the epidermis of the feet. In addition, lots of other materials were videotaped exiting and entering the feet (see chapters 4, 9 and 11).

The pedal phenomena were more pronounced if the feet were slightly moist and were inside shoes just prior to the exposure to the vaporized cocaine. The videos of material spontaneously exiting the feet are all taken during or immediately after exposure to the fumes of the incinerated cocaine. The enhancement of the phenomenon in slightly moist feet was discovered by trial and error. The subject's feet were moist one day after stepping in puddle of water and there appeared to be more particles exiting the feet while the feet remained moist.

To test the hypothesis that the microscopic organism was recruiting epithelial cells in order to create some sort of extra-cellular matrix, green ink was applied to the soles of the feet at one point during the investigation. The subject was examined for changes in the dermis and eyes involving the color green.

Observations were recorded using a Sony HDD digital video recorder from January 2008 until December 30, 2008. The videos were analyzed using the Sony Picture Utility software supplied with the video camera. This software allowed the frame by frame analyses needed to determine the mobility characteristics of this rapidly moving micro organism. Consecutive Sony Picture Utility still frames are separated by 33 milliseconds. A small number of videos are available on a DVD included with the book along with the photographs contained within the book in full color.

Large amounts of material extracted from the subject were collected in plastic containers for future analysis.

3

Evidence not Recorded on Videotape

The interviews with the complainants were elected to be placed in the introduction however they could have just as easily been included with the observations. The decision to place them in the introduction was made because most of them occurred prior to the decision to commencing this investigation. The other problem was that signed consents to reproduce and publish the informants' complaints were not obtained and therefore they must remain anonymous.

The observations were for the most part video taped, samples collected for future analysis and slides of samples were made. However, some of the videotaped data as well as all the collected samples and permanent slides made before April 2008 were deliberately discarded by the researcher's spouse. Despite the lack of physical evidence, it was decided to include the observations made by the research subject and the descriptions of the discarded samples.

The research project commenced in late August 2007 when the research subject began consuming between 600 mg and 1000 mg of vaporized cocaine daily between the hours of 9pm and 3am. The cocaine was smoked in the basement of his domicile where the subject also slept on a cot with a white cotton blanket. About two to three weeks later, the subject noticed that minute white objects were entering and exiting the ceiling halogen pot lights. At the time the white objects were thought to be baby spiders and not related to the presence of cocaine. At the time, the subject was expecting that the organism, if any did indeed exist, would be in the form of a macroscopic parasite – the so-called "coca mite".

About this time the subject began to experience paroxysms of paradoxical pruritis especially immediately after vaporizing the cocaine. The subject's feline also began experiencing the same symptoms. Sometimes the subject's cat would sit down and begin gnawing various parts of his body especially just superior to the dorsal base of his tail. A rash developed on the subject's left ankle and when he showed it to a co-worker she advised him that something appeared to be moving through his lower leg soleus muscle.

An emergency physician was consulted and he suggested that the symptoms were probably consistent with a subcutaneous vermis. He suggested taking vermox

100 mg twice daily for 3 days and undergoing a course of permethrin 5% dermal cream in case it was scabies. The advice was followed and vermox treatment started the next day and the permethrin treatment the next evening. The next day nothing abnormal was detected. The permethrin treatment and vermox were continued that day. Again, on the third day nothing abnormal was detected. That day, the vermox treatment was continued and instead of permethrin, a forty-five minute soak from the neck down in a bathtub full of hot water with two cups of dissolved Sunlight brand powdered laundry detergent was performed. Soaking in aqueous Sunlight laundry detergent was thought to be an alternative treatment for scabies.

The next afternoon and over the next approximately 24 hours dozens of C-shaped objects began to emerge from the dermis over the whole body from the neck down. They were approximately 5 mm in total length but were only 2 mm long because they were all folded in half resembling minute letters C. These objects appeared to be composed of dozens of aligned white stiff fibers. These C-shaped objects were saved using pieces of cellulose tape however they were all eventually discarded by the subject's spouse along with all the microscope slides and a jar of samples.

The subject's version of the origin of the C-shaped objects was met with disbelief and attributed to hallucinations by the reviewers. It was at this point that it was decided to purchase a video camera in order to record further phenomena if they were to occur. Unfortunately, the phenomenon of the C-shaped objects emerging from the dermis ended by the time a camera was procured and has not reoccurred.

During the first six months of video-taping only one phenomenon was recorded and retained (see video 37). All of the videotapes of the feline's symptoms and all but one video of the human subject were eventually erased when the 30 gigabyte hard drive of the camera was full. After six months, an external hard drive was purchased to store the rest of the videos and these videos make up the vast majority of the videos described in this book.

The majority of the videos that were erased showed essentially one phenomenon: specific longitudinal bands of epidermis located on the plantar surface of the feet could be removed painlessly and then seemingly regenerate within seconds. The painless character of the epidermal removal was not a consequence of the normal inhalation of vaporized cocaine; rather it appears to be a direct result of the presence of an inter-epidermal substance. The evidence that the lack of pain wasn't a result of the generalized analgesic effects of cocaine ingestion was that the analgesia was localized. Indeed, the removal of the epidermis anywhere else except in specific localized areas resulted in severe pain. Whether the analgesia was a result of the intradermal keratin-microorganism matrix or the analgesic effects of cocaine contained within the matrix or a combination of both is unknown.

The bands in the plantar surface of the feet began at the calcaneus, traveled distally parallel to and within three centimeters of the lateral border of the foot to about the proximal metatarsal heads. At this juncture, the band would move medially to the area between the proximal heads of the first and second metatarsals when the band would begin to move distal again for a couple of centimeters. At this point between the band would move laterally again to about the middle of the foot and would head distally again. The band was at this time located at the proximal border of the metatarsal pad over the third metatarsal. From here, the bands would move distally to the lower phalanges right to the toenails. At the toenails the bands would travel proximally along the dorsal surface of the foot and continue on towards the scalp.

The videos depicting the bands were sacrificed for two reasons. The first reason was practical. In the early stages of the research there was nowhere to store the data except in the hard drive of the camera. Admittedly, it was laziness, pure and simple. I was working seven days a week, eleven hours a day, in my clinic seeing patients and filming between four and six hours per day after work. The simple act of reading the instructions provided with the camera, installing the software to download the videos and purchasing more memory to store the videos was not performed until more than three months into the video taping or seven months into the research. Until then, less convincing videos were simply erased from the camera's thirty gigabyte hard drive when it became full in order to provide space for more compelling evidence.

The second reason was an extension of the first reason. I was sacrificing the less compelling videos out of laziness and it so happened that I decided that videos depicting the removal of epidermal bands with a pair of needle forceps was not compelling enough evidence. Indeed, I reviewed the films with a physician colleague and he stated that all he could see was human mutilation.

Admittedly, the evidence of the bands in the videos was extremely subtle. The evidence consisted of the removal of the band as it moved distally across the metatarsal pads. The band would regenerate unless all the epidermis was removed. If all the epidermis was removed then the band would simply move laterally to the next strip of epidermis. The appearance of the regeneration of the band was subtle: imagine slowly filling a collapsible garden hose with water. Basically, the thinned out dermis became slightly thicker, not exactly compelling evidence of a parasitic worm. I refer to a worm here because that was what I was expecting to find, a worm in the dermis. After the initial samples were determined to be epithelium (see figures 9.12 and 9.13), the methodology of epidermis removal was abandoned in favor of focal material sampling and passive video observation.

The bands were not only located on the feet. The bands were also observed to be located peri-orally extending between the oral labial angle and the nare along both sides of both nasal-labial folds. These folds would become prominent while ingesting cocaine - the left side more than the right - giving the face a distorted and asymmetrical appearance previously attributed to involuntary muscle contractions resulting from the neurological effects of cocaine ingestion.

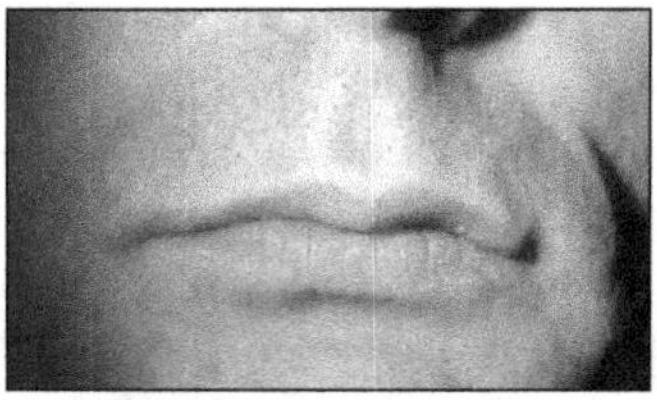

Fig. 3.1 This photograph depicts the typical facial distortion experienced by the subject after ingesting cocaine. The fullness on the left side of the face was probably the result of an intra-epidermal keratin-microorganism matrix that traveled through the subject. There is a white particle on the epidermis located about 15 mm above the upper lip on the left side. The left side of the upper lip appears swollen and the inferior aspect of the lip is white indicating the possible presence of the microorganism.

The other category of video evidence that was not retained was the ocular changes observed in a cat. The subject's feline began to demonstrate paradoxical and paradoxical ocular tearing combined with changes to the pupils and iris. The cat was suspected of consuming cocaine crumbs on occasion. The camera was placed by the cat's bed and the eyes videotaped after vaporized cocaine was blown into its face. White particles similar to the white particles seen in figure 6.7 were seen exiting its eyes. The cat's iris became irregular as though there was another iris was located underneath his normal iris. The videos also showed that the cat's upper lip became noticeably swollen after ingesting cocaine mixed into its wet food. This feature is analogous to the facial distortion seen in the human subject (see figure 3.1). These videos were also sacrificed in the purges of the camera's memory.

These videos were deemed less compelling for two reasons. It was decided that evidence from humans was more important and expertise in cat physiology was not readily available. Again, the same physician colleague attributed the ocular changes to feline peripheral cataracts after viewing the videos. At this point it was decided to focus on collecting human evidence.

With the initial skepticism taken seriously, it was decided only to retain more convincing evidence. Even though the evidence is referred to as more convincing, it is acknowledged that there will always be credible criticism of the argument made here that this constitutes evidence of a new micro organism. Therefore, due to the anticipated controversial nature associated with any research of this nature, hundreds of videos were compiled to corroborate the descriptions detailed in this documentation of the investigation. Each video was assigned a number by the camera and the selected videos will be referred to by these same numbers preceded by. The photographs are taken from still frames of the videos produced by the Sony Picture Utility software and color versions of these photographs are included on the DVD accompanying the book. All of the videos described in the book are included in the DVD that is provided with the book.

4

The Phenomenon of the Disappearing Hair Follicles

The paradoxical paroxysmal appearance of hair, mainly pubic hair, stuck on the feet, was seen as an indication that something abnormal was occurring involving the subject's hair and also involving the feet. Although it was one of the first observations, it was regarded as normal. This phenomenon was an observation that took several months to recognize as abnormal because it is not abnormal to see the occasional hair stuck to the bottom of a foot.

There were several abnormal characteristics regarding the phenomenon of the hair on the feet that eventually lead to including it in the research. Until this time, the hair was brushed off the feet with forceps and the subject's hands after blowing cocaine vapors onto the feet and prior to extracting material from the feet with the forceps.

The hair was noted to be there immediately after removing the socks indicating that the hair may or may not be entering the socks from within the feet. The subject did not ambulate barefoot, donning socks immediately after bathing in the morning, and therefore, there was little or no opportunity for hair to become stuck to the feet. It should be stated here that this observation was made in retrospect after it was determined that hair was indeed entering the dermis. By extension, once hair was determined to be entering the subject through the epidermis, it was surmised that the hair located on the feet was arriving by way of exiting the dermis.

The observation that eventually led to the phenomenon being included in the research was that there seemed to be much more pubic hair on the feet when the pubic regions were pruritic. The connection between the appearance of pubic hair located on the feet after relieving the pruritis with scratching and a new phenomenon was made after a rather intense episode of pubic pruritis. After that episode of pruritis, there were significantly more pubic hairs seen on the plantar surface of the feet than usual. It was then decided to undergo a simple experiment.

On the assumption that somehow stimulating the pubic hair was causing it to enter the dermis either through the fingers or the pubic epidermis, it was surmised that perhaps stimulating the hair on the feet could produce the same effect. Many videos were recorded demonstrating this phenomenon to show that the phenomenon was not a figment of the imagination. The results of the experiment are detailed below.

Video 1 This video shows a hair follicle entering the toe then exiting the toe and then disappearing into the foot. The hair disappears into the 5th digit then reappears for one frame and then disappears again.

Video 2 This video shows a hair entering the toe. Prior to disappearing, the arc of the hair straightens slightly without any stimulation. A consecutive frame by frame analysis is shown below.

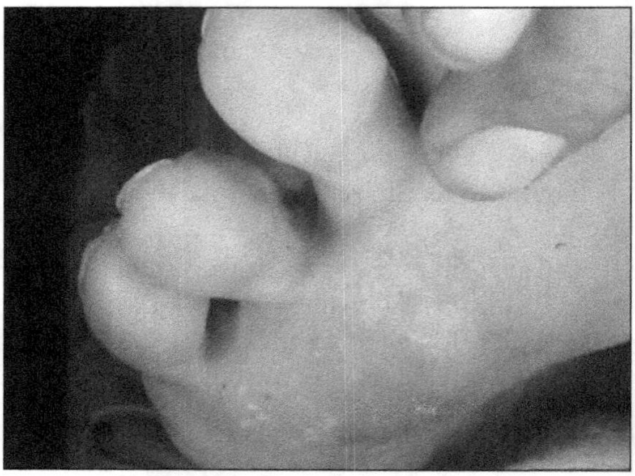

Fig. 4.1 This photograph shows the hair follicle as seen prior to it entering the epidermis.

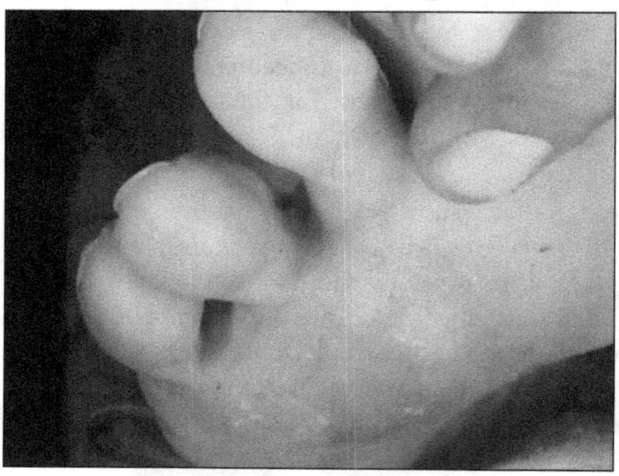

Fig. 4.2 This photograph shows the same hair follicle 33 ms after Figure 4.1. Without stimulation, the curvature of the hair follicle has lessened.

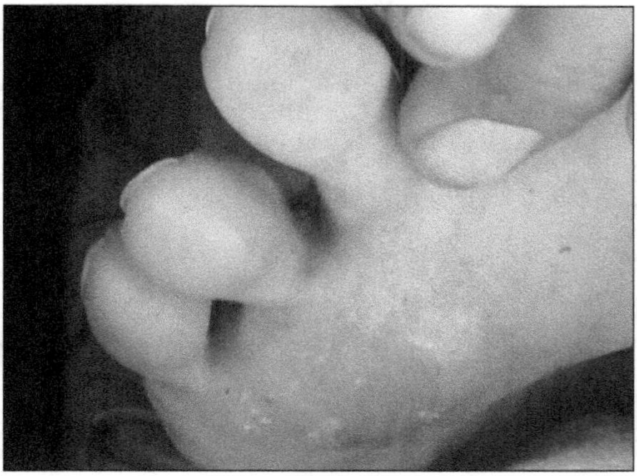

Fig. 4.3 This photograph, taken 33 ms after Figure 4.2, shows that the hair follicle has disappeared.

Video 3 This video shows a hair disappearing from the plantar surface of the right fourth toe. There is some white material visible beneath the second, third, fourth and fifth toenails.

Video 4 This video shows a hair disappearing from the left third toe. There is white material seen beneath the second, third, fourth and fifth toenails.

Video 5 This video shows a hair entering the toe. A consecutive frame by frame analysis shows the hair just beneath the epidermis.

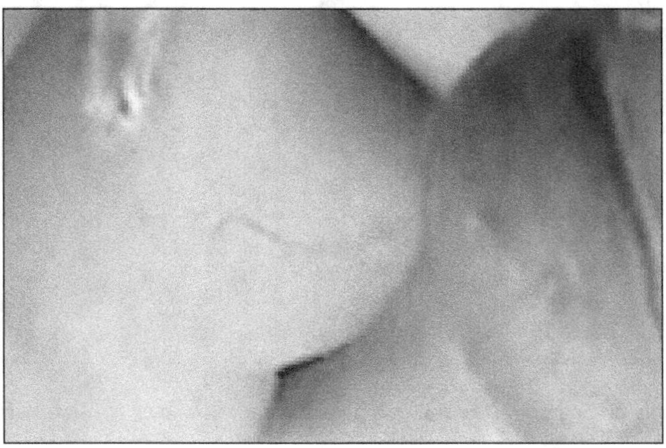

Fig. 4.4 This photograph depicts a hair follicle on the lower right fifth digit prior to contacting the left thumb.

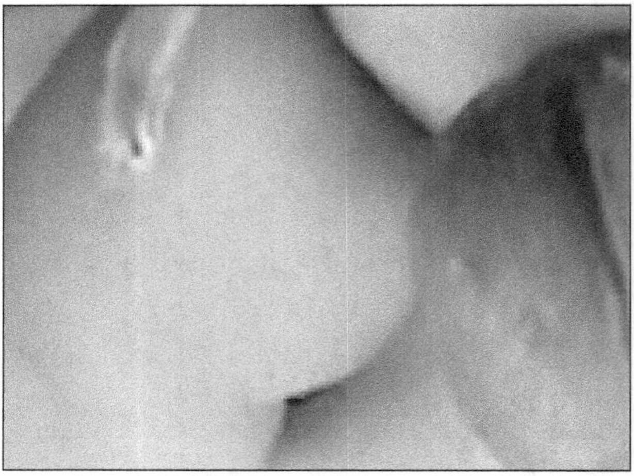

Fig. 4.5 This frame shows the same hair follicle as Figure 4.4 as contact is made with the left thumb 33 ms later. The hair is still partially and faintly visible. A small amount of subunguinal white material is seen on the fifth toe.

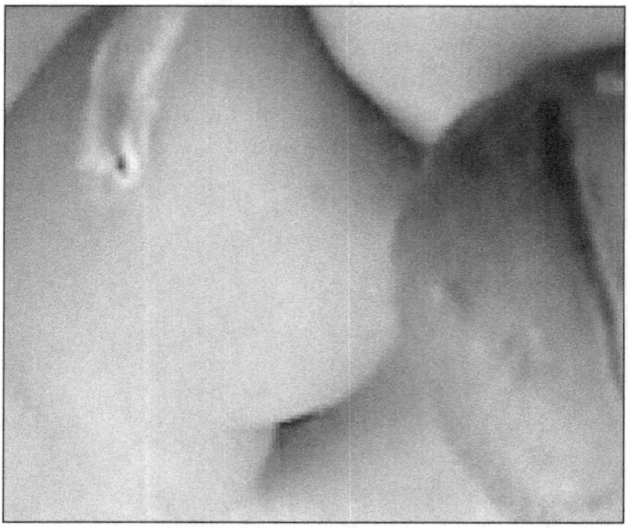

Fig. 4.6 This photograph, taken 33 ms after Figure 4.5, shows that the hair is no longer visible.

Video 6 This video shows a hair disappear into the plantar surface of the right foot.

Video 7 This video shows a hair disappear into the plantar surface of the right foot after stimulation by the left first digit.

Video 8 This video shows a hair disappear into the plantar surface of the left foot after being stimulated by the left first digit.

Video 9 This video shows a hair from the right foot slowly disappear into the left thumb.

Fig. 4.7 This photograph shows three phenomena: a hair can be seen partially imbedded in the epidermis of the thumb pad (a semi circular faint line starting at the mid cuticle and ending at the lateral aspect of the thumb), and; white particles are seen on the green-inked epidermis. The ink was applied to determine what role, if any, the epidermis plays in the structure of the microscopic organism. Figures 7.6, 8.2 and 8.3 all show a green colored substance present in the orbital and peri-orbital regions indicating that the epidermis from the feet makes its way to the eyes.

Video 10 This video shows an extended version of the above video.

Video 11 This video shows a hair disappear into the left foot after stimulation from the left thumb.

Video 12 This video shows a hair slowly disappearing into the thumb. Figures 4.8 to 4.14 are taken consecutively over a span of 14.03 seconds.

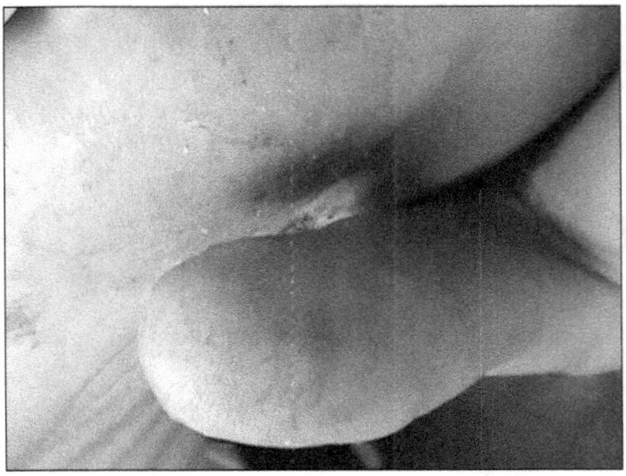

Fig. 4.8 A hair attached to the foot as seen immediately after contact with the thumb pad.

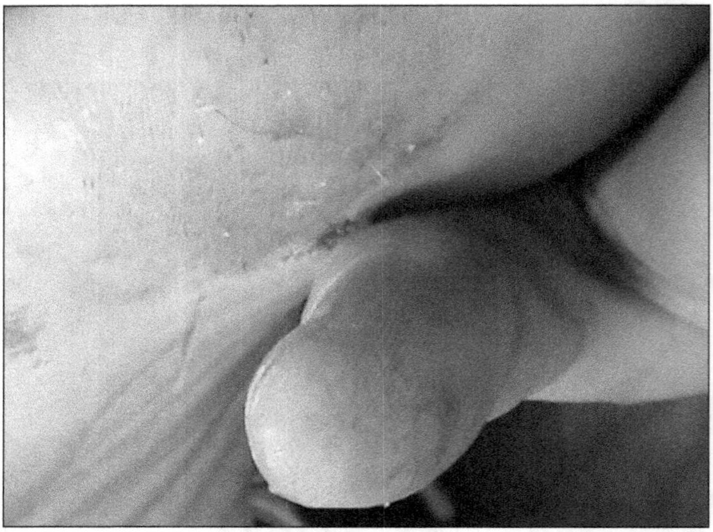

Fig. 4.9 Seconds later, the hair follicle is almost entirely imbedded in the thumb pad.

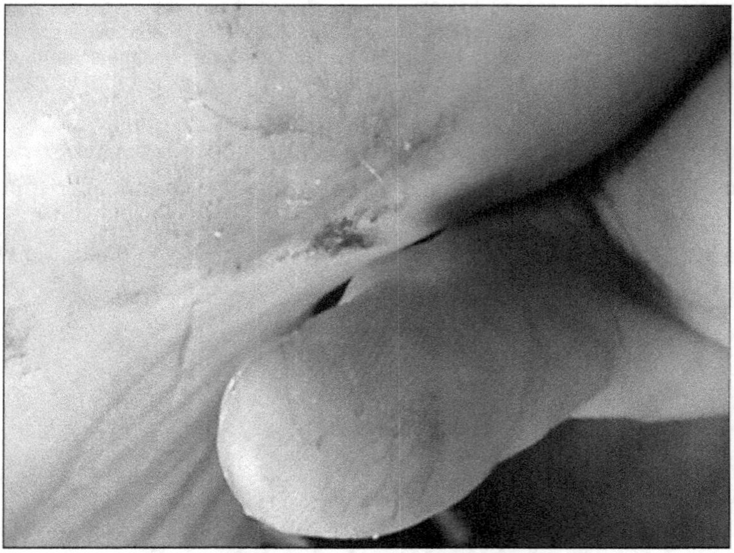

Fig. 4.10 The hair follicle appears to be re-emerging from within the thumb pad.

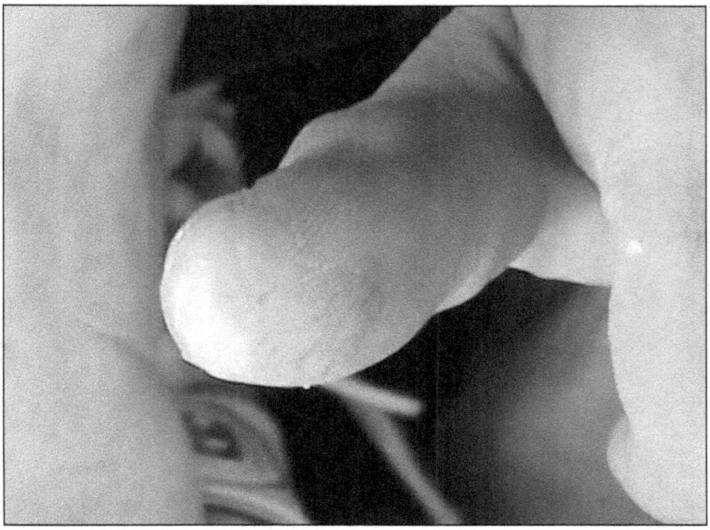

Fig. 4.11 The strand appears a little fainter and appears to have sliced through a small part of the nail.

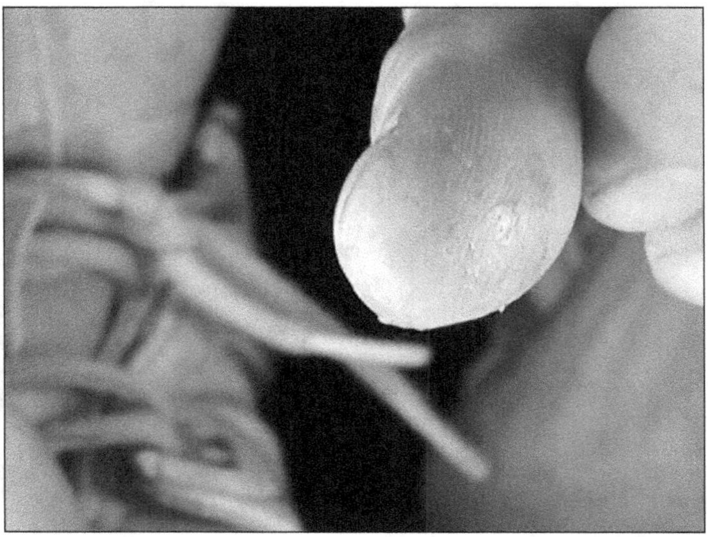

Fig. 4.12 The hair follicle appears fainter. Note that the hair follicle is protruding perpendicularly from surface of the nail.

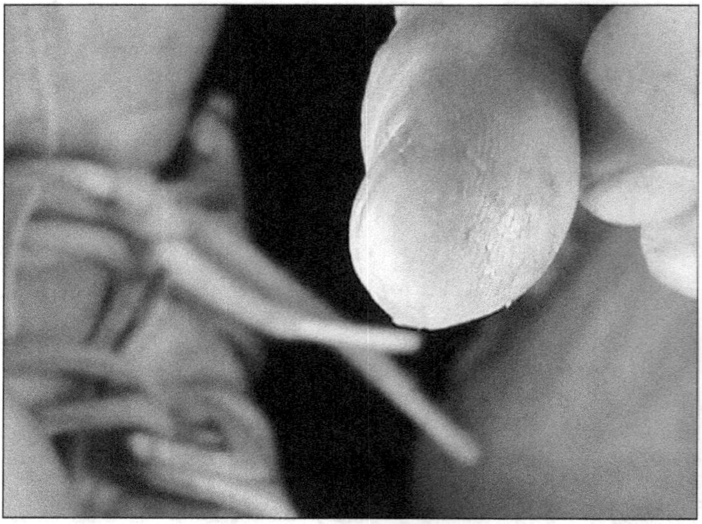

Fig. 4.13 The hair follicle is fainter and again the hair follicle appears to have sliced through the nail.

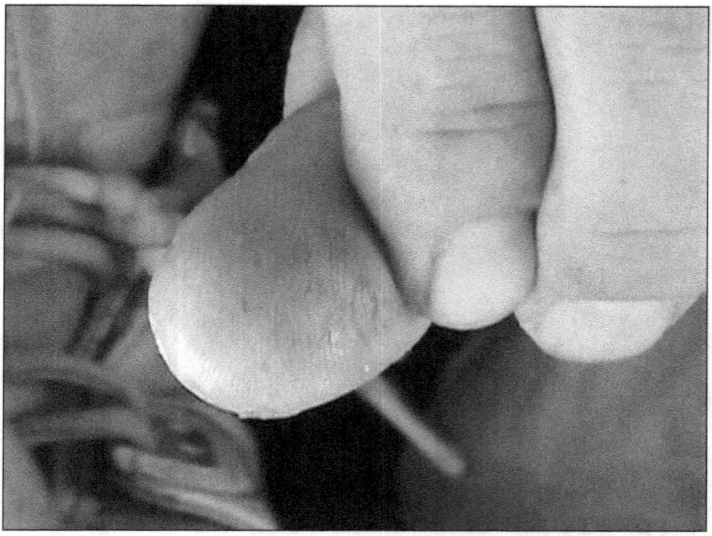

Fig. 4.14 In this photograph, the hair follicle appears a bit darker. A close study of the hair demonstrates that the hair follicle is outside the dermis laterally. As the hair is examined medially it appears to become intraepithelial and gradually becomes fainter. The hair follicle is only visible between the ridges of the thumbprint indicating that the hair follicle is in the epidermis. The hair follicle eventually disappears completely after it is compressed between the first and the second digits.

Video 13 This video shows a hair entering the fifth lower digit after stimulation by the left thumb and reappearing for one frame between the toe and the foot and then entering the distal metatarsal pad.

Video 14 This video shows a hair entering the left foot or left thumb. The disappearance of the hair was so rapid that even a frame by frame analysis could not discern whether the hair entered the thumb or foot. What is seen is that the hair is present in one frame and then is completely disappeared in the next frame taken 33 ms later.

Video 15 This video shows a hair entering the right foot or left thumb. Again, as seen in video 14, the velocity of entry was so fast that it is impossible to determine whether the hair disappeared into the thumb or foot.

Video 16 This video shows hair disappearing into fingers. The first hair disappears after being taken from the right foot and then is compressed between the left first and second fingers. The second hair disappears from the left foot after it is stimulated by the second left finger.

Video 17 This video shows a hair entering a finger and then exiting the same finger. The hair appears to have lost its previous curvature and is completely straight after exiting the finger.

Video 18 This video shows a hair follicle disappearing into the second toe. The hair is seen at the beginning of the video and then disappears without stimulation and then reappears. The hair disappears and reappears several times before finally disappearing permanently when it is stimulated by a pair of forceps. The feet are grossly swollen as happens when the toes are compressed together. The following 16 photographs are sequential but are not consecutive and are all taken within a 1 minute 21 second and 240 millisecond timeframe.

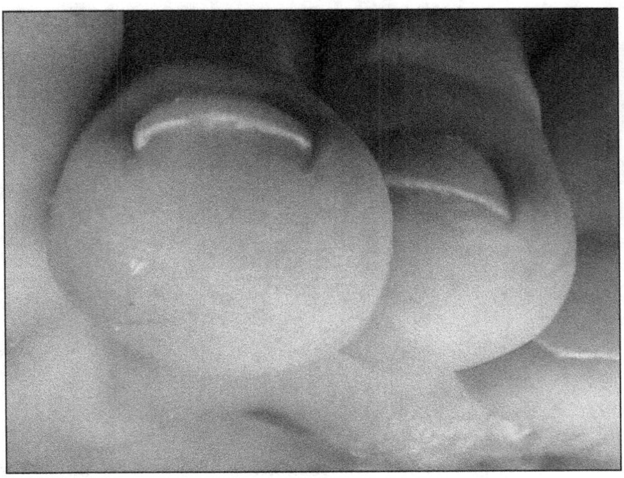

Fig. 4.15 This frame shows a hair located on the second toe.

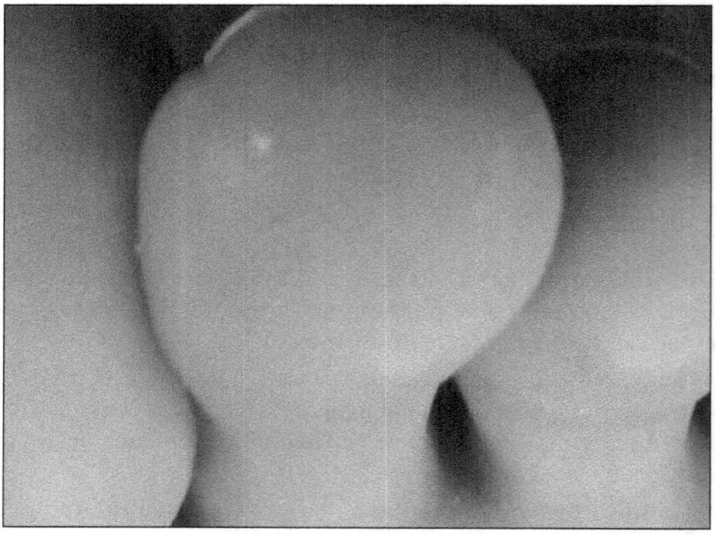

Fig. 4.16 This frame, taken 32.13 seconds later, shows that the hair is virtually invisible. The area of the toe where the hair was visible previously is still in focus as is evidenced by the visibility of the ridges of the toe print. It is surmised that the hair has temporarily entered the epidermis.

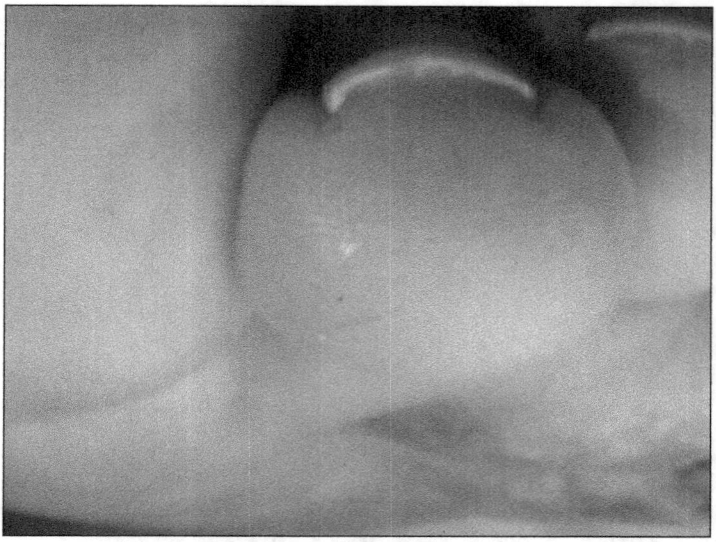

Fig. 4.17 In this photograph, taken 5.13 seconds after Figure 4.16, the hair is again visible on the surface of the epidermis.

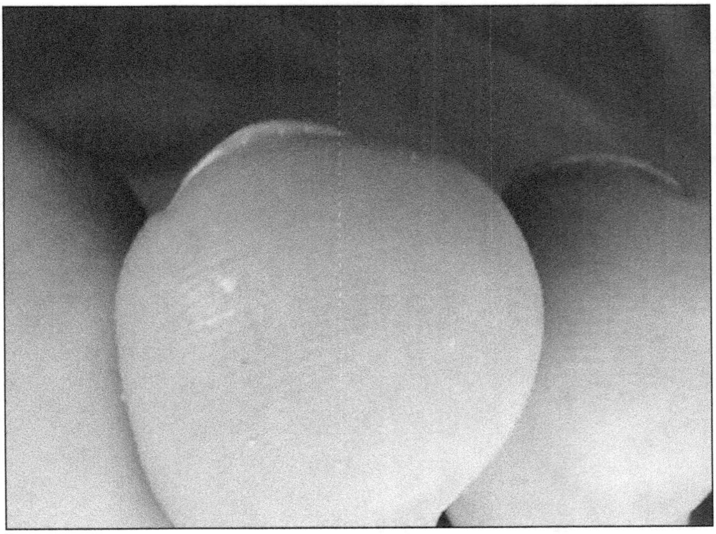

Fig. 4.18 This photograph shows that the hair is again within the epidermis, 4.36 seconds after Figure 4.17. The area of interest is in focus, again evidenced by the visibility of the toe print ridges.

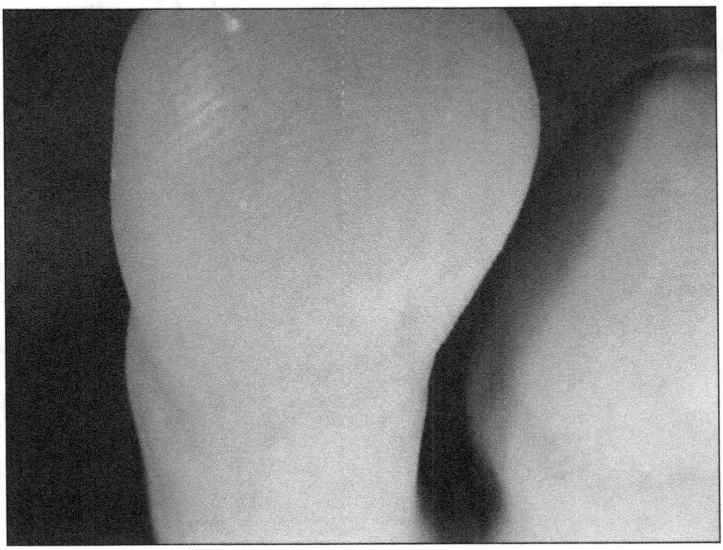

Fig. 4.19 This photograph, taken 6.03 seconds after the previous figure, shows the hair reappearing for the second time.

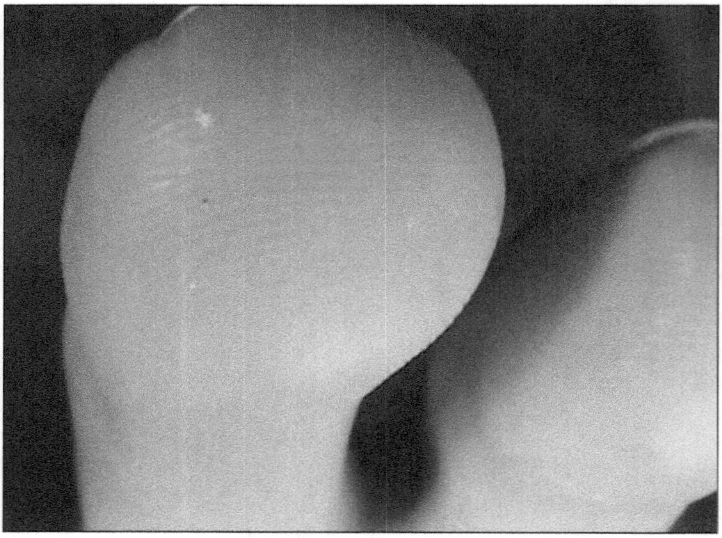

Fig. 4.20 This photograph shows the hair is only partially and faintly visible 5.03 seconds later. As before, the focus can be verified by the visibility of the toe print ridges.

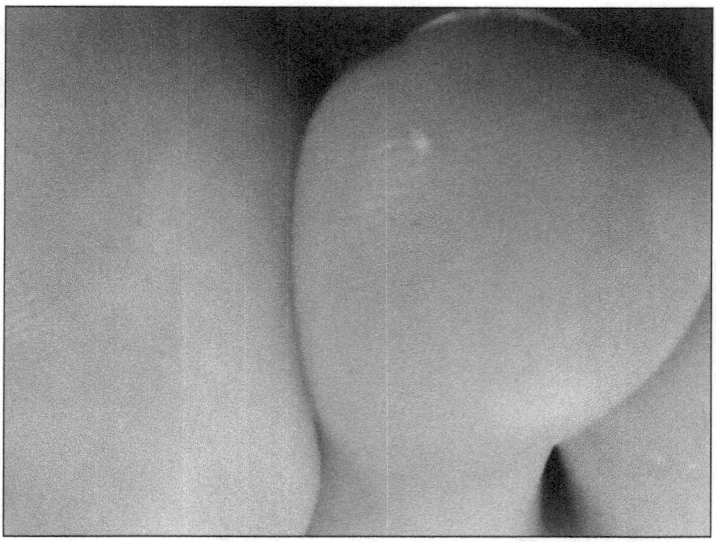

Fig. 4.21 This photograph shows that the hair is virtually invisible 15.34 seconds later.

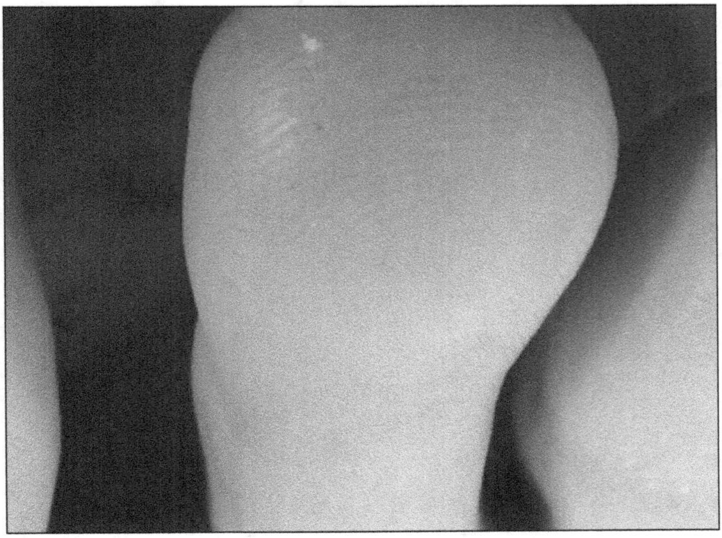

Fig. 4.22 This frame shows that the hair is only partially visible 3.9 seconds after Figure 4.21.

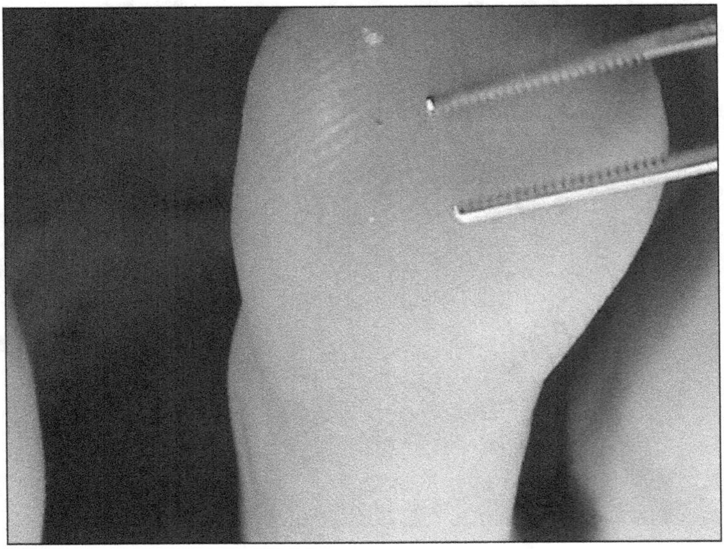

Fig. 4.23 This photograph, taken 2.4 seconds after the photograph of the hair in Figure 4.22, shows that sections of the hair follicle not visible on the epidermis, while other sections are visible on the surface. The hair will disappear after stimulation with the forceps as demonstrated in the next 4 figures.

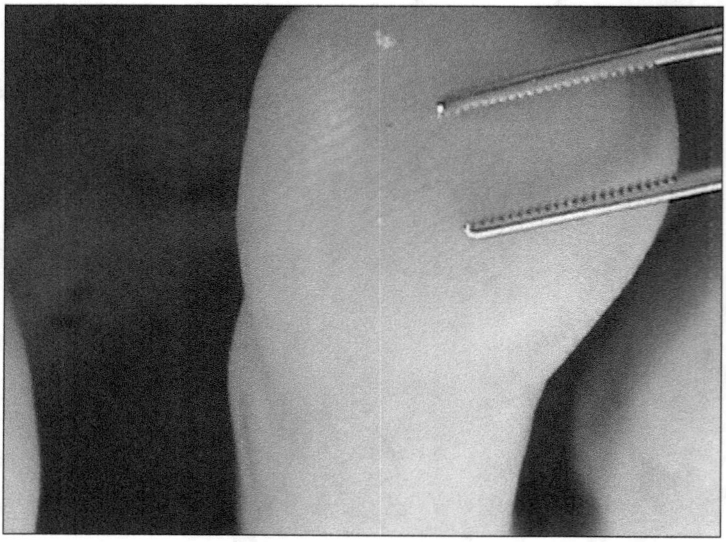

Fig. 4.24 This photograph shows virtually no change after another 0.15 seconds.

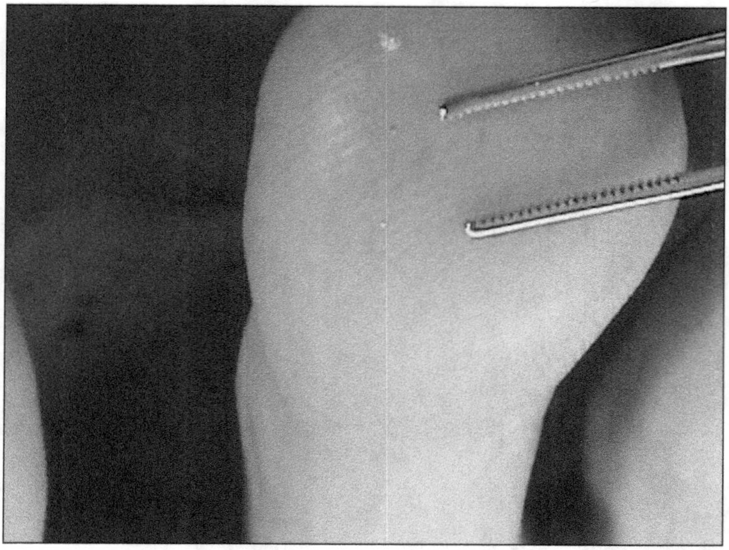

Fig. 4.25 This photograph taken 0.23 seconds after the photograph in Figure 4.24 shows virtually no change in the visibility of the hair.

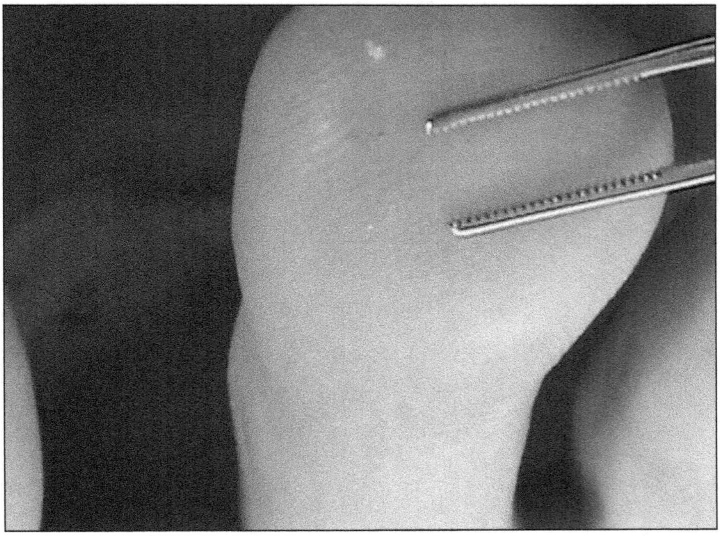

Fig. 4.26 This photograph is taken .03 seconds after the previous figure and shows virtually no change in the visibility of the hair.

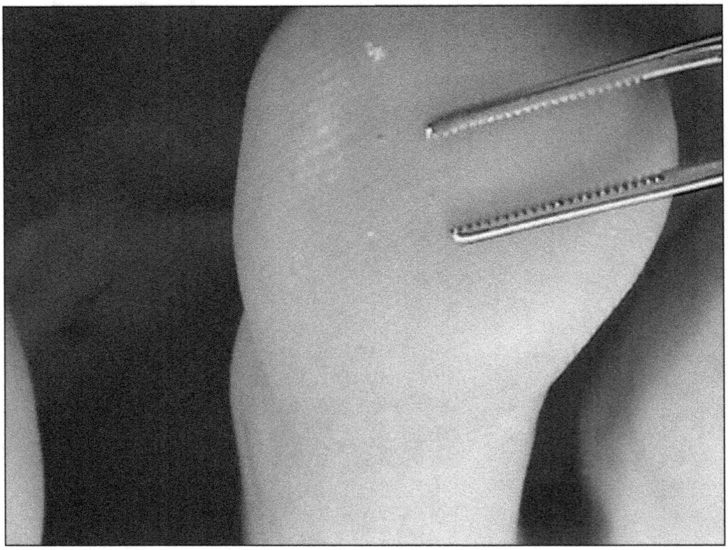

Fig. 4.27 This photograph shows that all sections of the entire hair disappears permanently within 0.03 seconds. The observation that the entire hair disappears at the exact same moment supports the idea that the separate sections of visible hair seen in Figure 4.26 are all part of the same hair follicle as seen in Figure 4.15. The lighting and focus are identical to the previous photograph.

5

Particles Entering the Eyes

After several months of researching the feet it was decided to include the eyes to determine whether the ocular changes in the cat were reproduced in the human subject. The camera was set up on a stand and trained on the eyes immediately after ingesting cocaine. Lighting was varied occasionally using a 300W Welch-Allyn examination lamp as well as a Welch-Allyn headlight held in the left hand.

The results were surprising. While anticipating observing small white particles exiting the eyes as observed while filming the cat's eyes (see chapter 6), it was not expected to observe hair and linear particles entering the eyes. In addition, the medial portions of the eyes were often covered in plaques of material and the irises appeared to be protruding through the cornea (these phenomena are covered in chapters 7 and 8). The observation of particles and hair entering the eyes, recorded on video, are compelling evidence that a new micro organism exits.

Video 19 This video demonstrates several interesting phenomena. The first and most significant is that it shows two thin perfectly straight particles entering the globe. A frame by frame analysis confirms the same. Another phenomenon seen in this video is that the eyelashes are heavily coated in a white material.

A sagital band of material is seen extending from the upper lid to the lower lid along the medial border of the globe which is partially covering the cornea, obscuring part of the iris. The iris appears to occupy, and thereby obliterate, the anterior chamber of the eye.[6]

Each of the following consecutive frames is separated by 33 ms.

[6]The normal structure of the eye includes an anterior chamber which is a space located between the iris and the cornea. This space is normally filled with vitreous humour. The anterior chamber is not visible in many of the photographs of the eye contained in this examination. Instead of an anterior chamber, the iris is seen abutting the cornea thereby obliterating it.

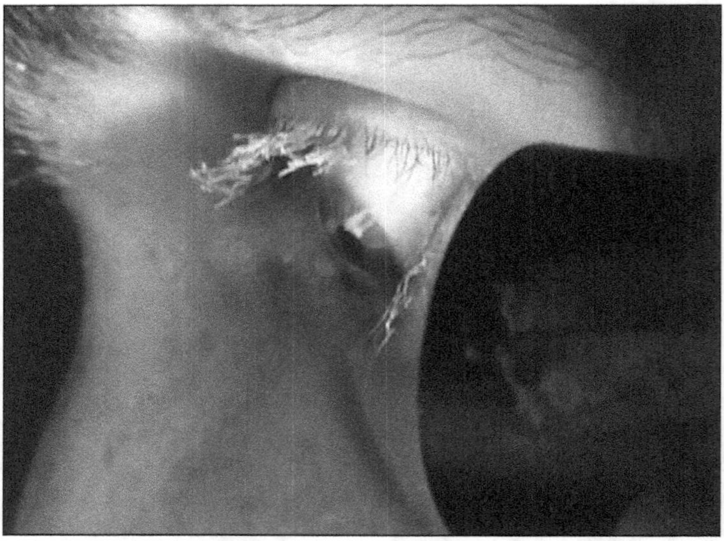

Fig. 5.1 A faint strand is barely visible inferior to the pupil. Also present in this photograph are eyelashes that are heavily coated in a white material and two white bands at the three o'clock position on the cornea, one of which terminates in luminescent white area.

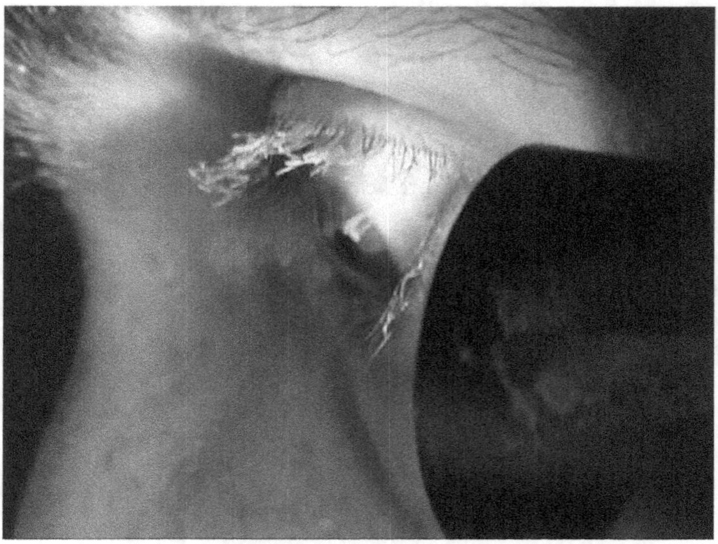

Fig. 5.2 The next frame in the video taken 33 ms later shows the strand is now more visible and is closer to the pupil.

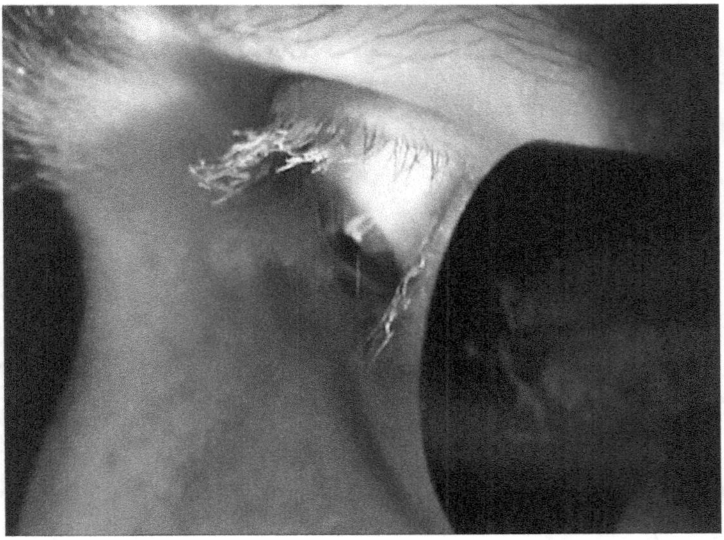

Fig. 5.3 The next frame shows the strand is about to enter the optic globe at the site where the white spot is located.

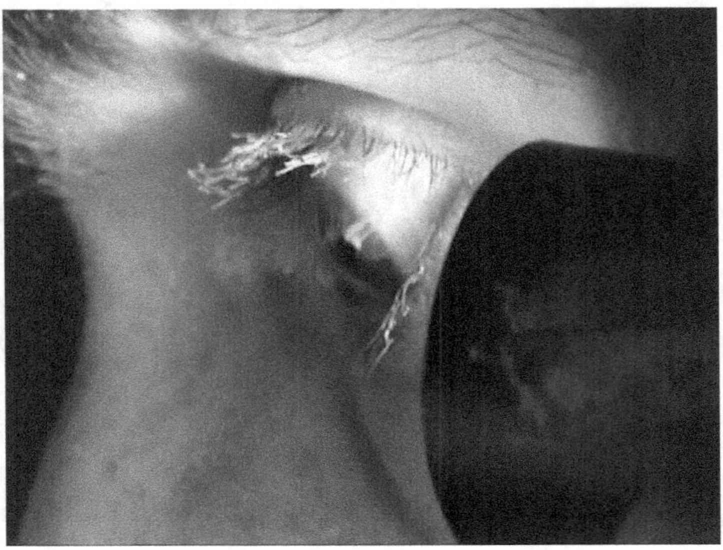

Fig. 5.4 The next frame shows the strand almost completely inside the optic globe. A small part of the strand remains visible at the 5 o'clock position. The strand appears to be entering the eye at a point of increased luminescence located at the three o`clock position on the inner border of the iris which may indicate a protruding white particle. Note that the iris appears to be occupying the anterior chamber supporting the conjecture that the area of increased luminescence is a particle protruding through the cornea from the iris.

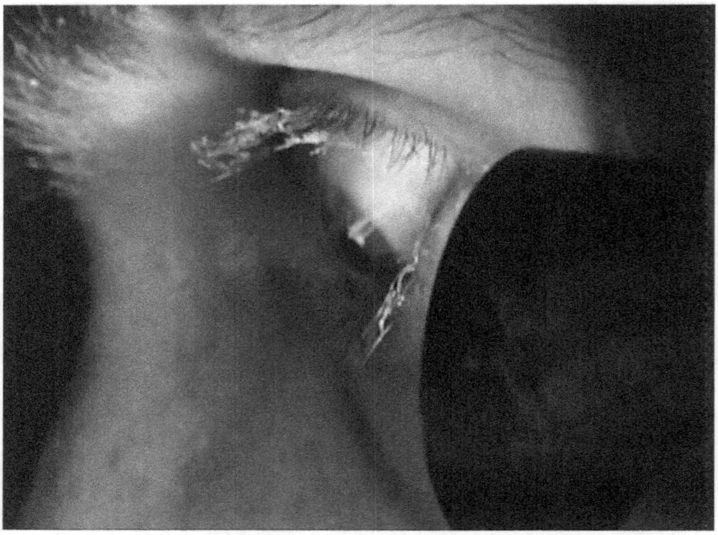

Fig. 5.5 Within a few seconds a second strand is seen heading towards the optic globe.

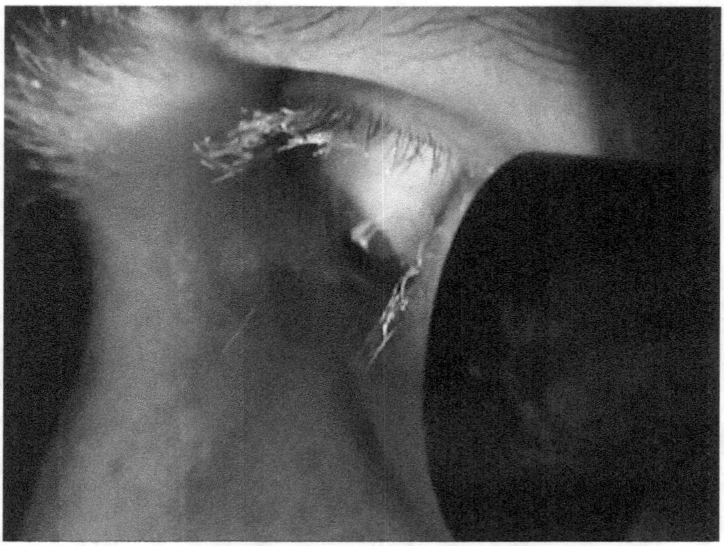

Fig. 5.6 The next frame shows the second strand closer to the globe.

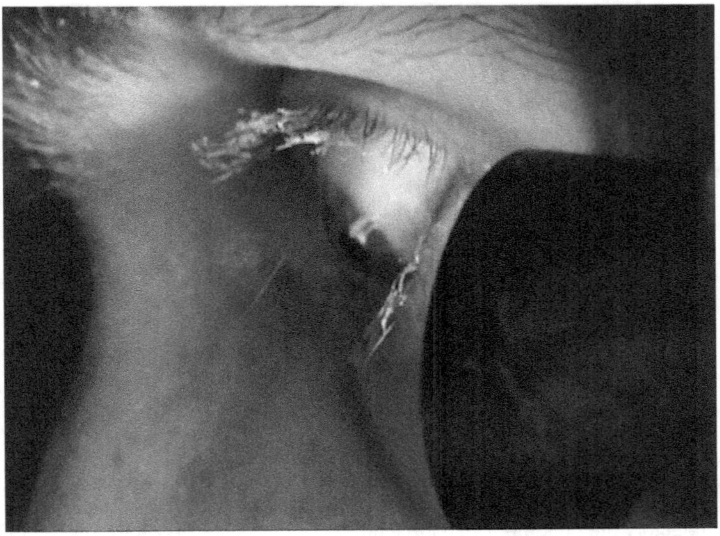

Fig. 5.7 The next frame shows the strand closer to the optic globe.

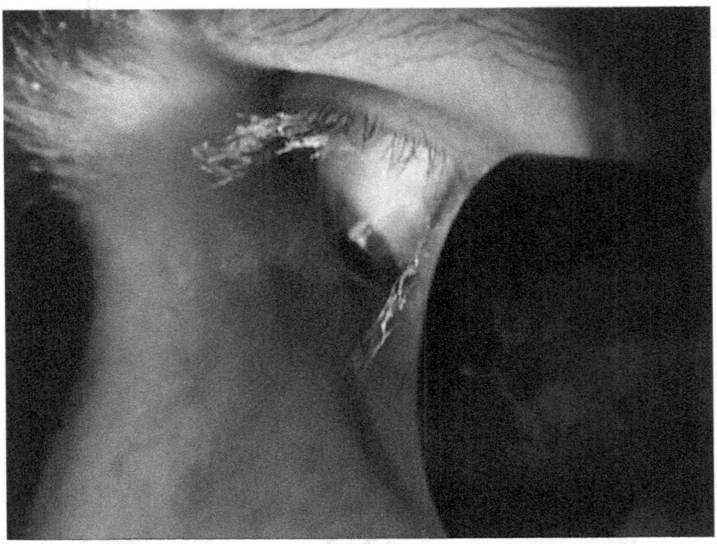

Fig. 5.8 This photograph shows the strand immediately prior to entry into the globe.

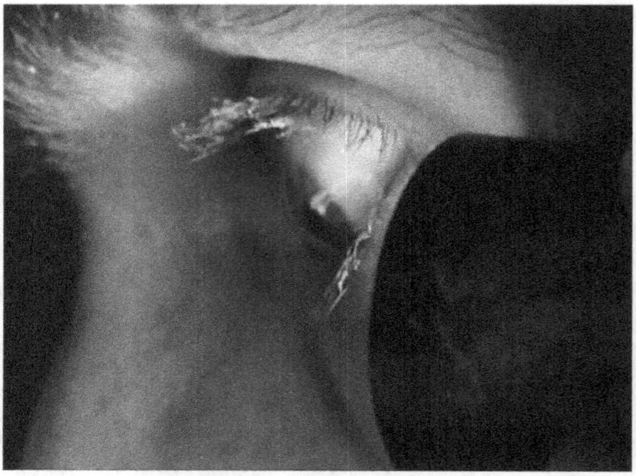

Fig. 5.9 The next frame, taken 10 ms later, shows that the strand has completely disappeared into the optic globe. As with other mobile particles seen in the other videos, the lines of trajectory are congruent with the long axes of the particles, in this case the particles are thin strands. The possibility that these are strands of lint propelled by air currents is highly unlikely. If these strands were strands of lint they would not be aligned in a perfectly straight line with no turbulent movements.

Video 20 This video shows a hair follicle coated in a white substance entering the eye.

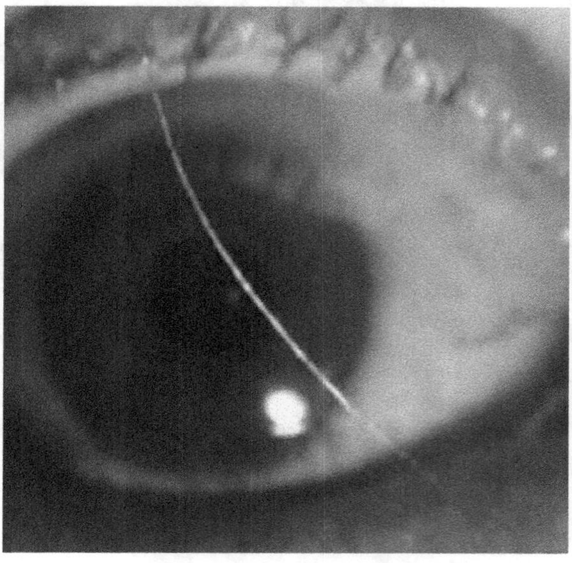

Fig. 5.10 A photograph of the hair follicle seconds before entering the eye. The entire length of the hair is covered in what appears to be dozens of white particles. This has the same appearance as the hair on the forearm (see Figure 13.1).

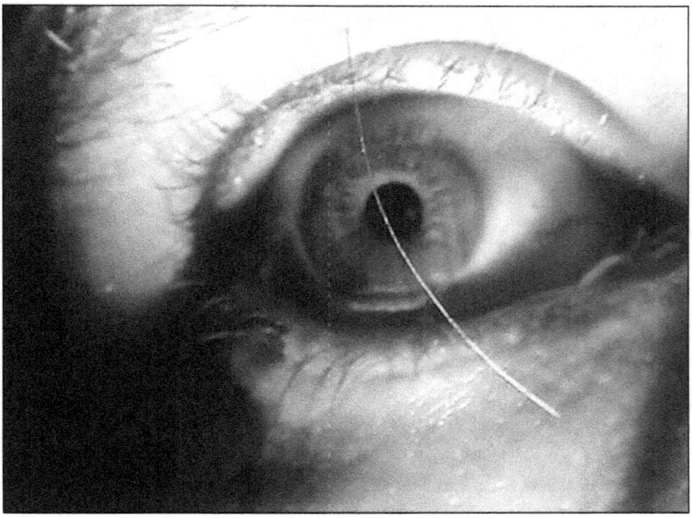

Fig. 5.11 This photograph shows another view of the hair follicle and the white particles lining it.

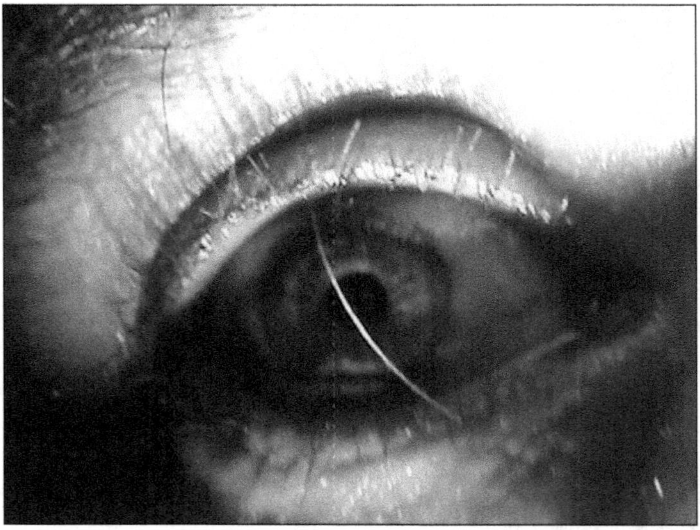

Fig. 5.12 This photograph shows the hair follicle as it begins to move towards the eye. The minute formerly discrete particles now appear to be confluent except at the very distal end. This could be the result of a rapid movement of the hair causing blurring of the image or it could be the result of movement among the individual particles into a confluent mass. The significant thing to note is that either the white material transforms itself or the white material does not disappear with movement. This becomes significant when combined with Figure 5.14 which shows the superior half of the white material is not visible therefore it must have entered the eye.

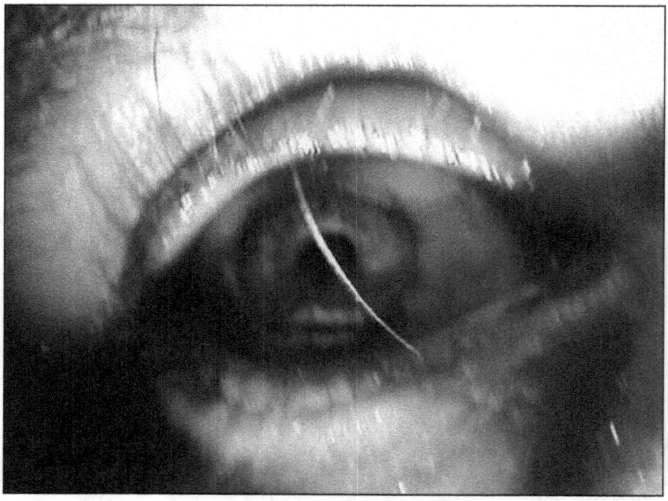

Fig. 5.13 A hair follicle is seen immediately prior to its entry into the cornea. Note that the entire length of the strand is reflecting light and or is luminescent.

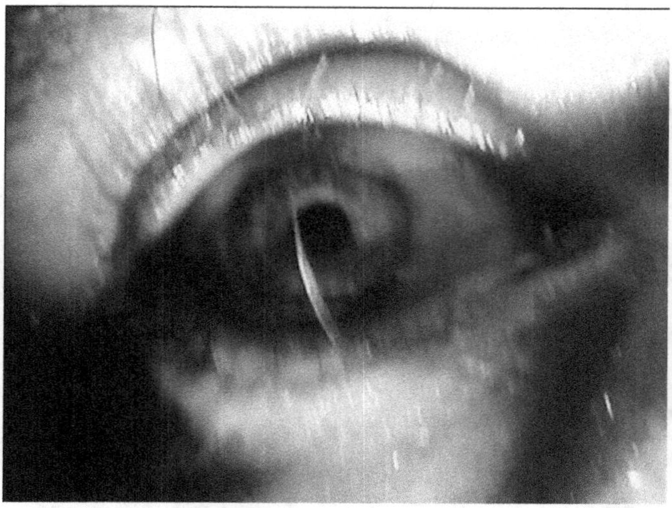

Fig. 5.14 The next frame of the video depicted here, taken 33 ms after Figure 5.13, shows the same hair follicle entering the cornea 33 ms later. The superior quarter section of the hair is not visible at all. The quarter inferior to the superior quarter is merely a brown line and no bright white reflection or luminescence is detectable. Figure 5.12 demonstrated that the white reflective or luminescent quality of the hair follicle does not disappear due to movement alone. There is no other explanation as to what happened to the white section of the hair other than it has entered the globe. Note that the distal tip of the hair has also disappeared again presumably into the lower eyelid. Therefore, the hair follicle coated with white material entered the eye by both ends first, followed by the middle portion of the hair.

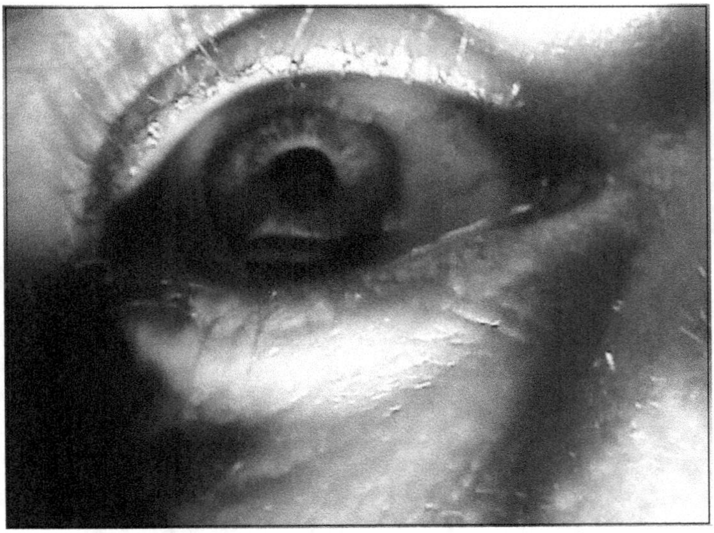

Fig. 5.15 The next frame shows no trace of the hair follicle. Note that in Figures 5.11 to 5.14 the iris appears misshapen and asymmetrical.

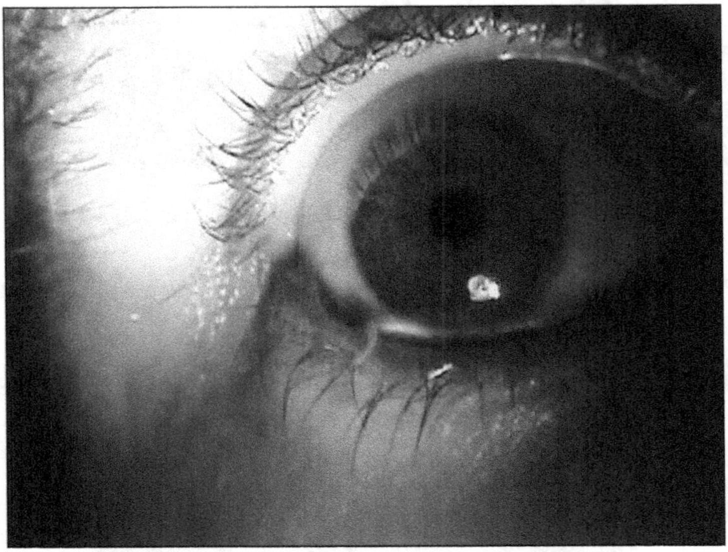

Fig. 5.16 A photograph of the eye taken later in the same video shows a peach colored material is seen exiting the optic globe and extending across the lower lateral lid margin.

Video 21 This video shows a particle entering the dermis inferior to the globe. Although this particle is not technically entering the optic globe, it was included here is the phenomenon was captured while filming the eye.

Figures 5.17 to 5.24 are consecutive frames that show a white particle flying through the air (from the bottom right corner) and then entering the dermis. The figure 5.24 shows the dermis 33 ms after the time of impact. An important observation is that the particle elongates as it approaches the dermis as is seen in figures 9.2 to 9.8 where the particle elongates after it exits the dermis. The particle line of trajectory is congruent with the long axis of the particle throughout the flight path indicating that the particle is unlikely to be a piece of dust or lint propelled by wind currents. This observation is not an isolated phenomenon and is reproduced in all the frame by frame sequences reproduced in this book (see Figures 5.1 to 5.9 and Figures 9.2 to 9.8).

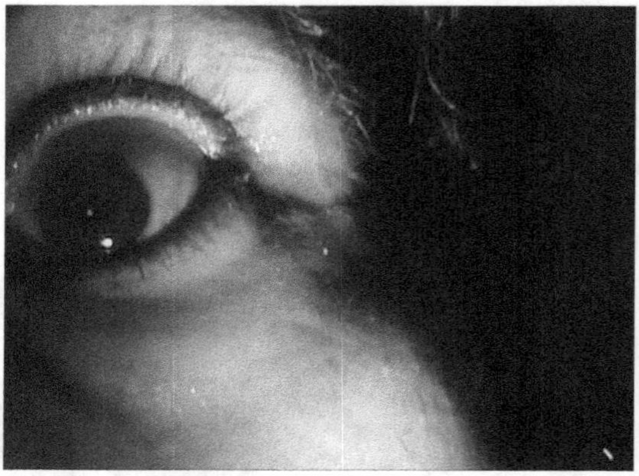

Fig. 5.17 A bright white particle is seen entering the video at the right lower corner.

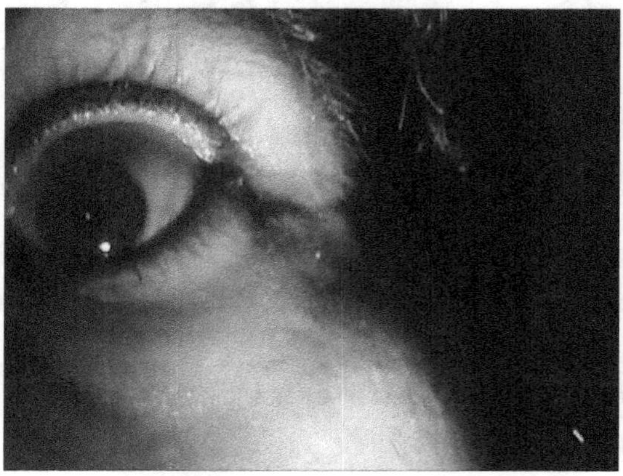

Fig. 5.18 The particle continues to ascend and appears to elongate slightly and is slightly less bright. The trajectory line and the long axis of the particle remain congruent.

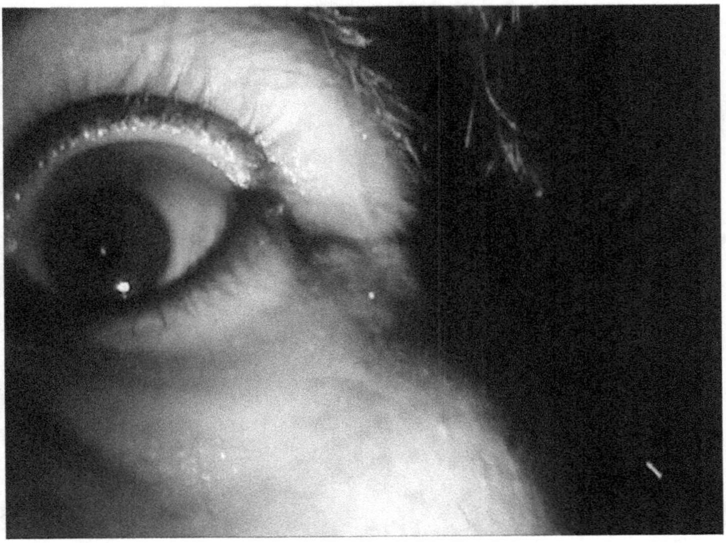

Fig. 5.19 The particle continues to ascend, elongate and darken. Again, the trajectory and long axis of the particle remain congruent.

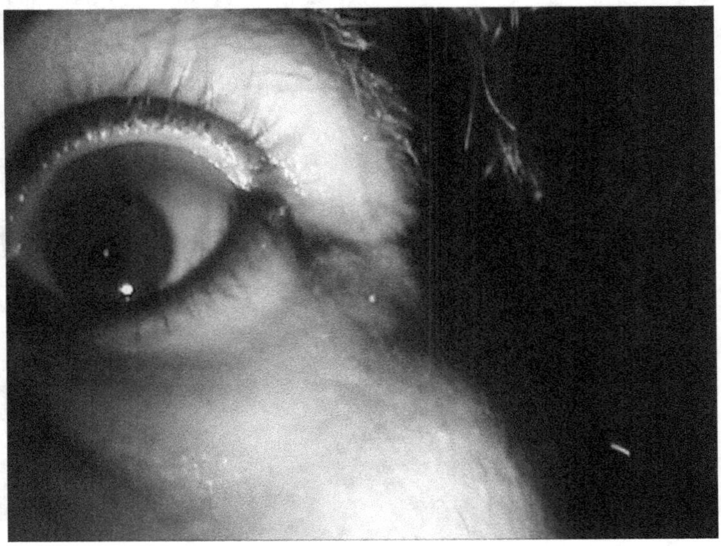

Fig. 5.20 The trajectory changes and so does the long axis, both remaining congruent.

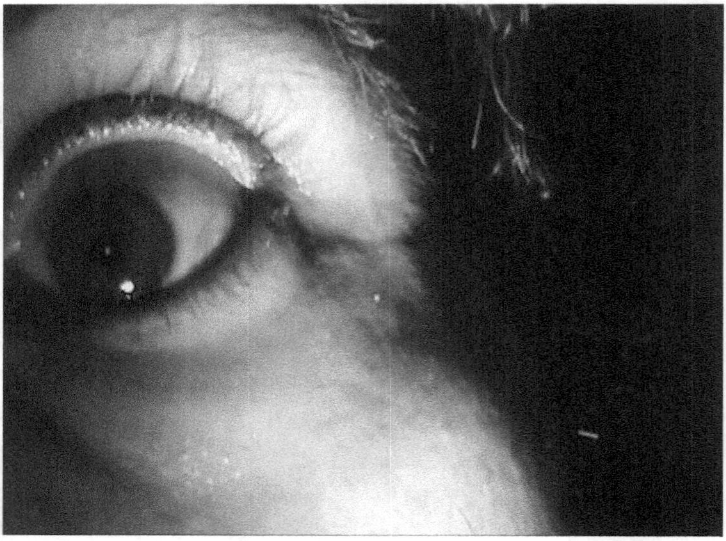

Fig. 5.21 This frame shows the particle continuing to move more horizontally.

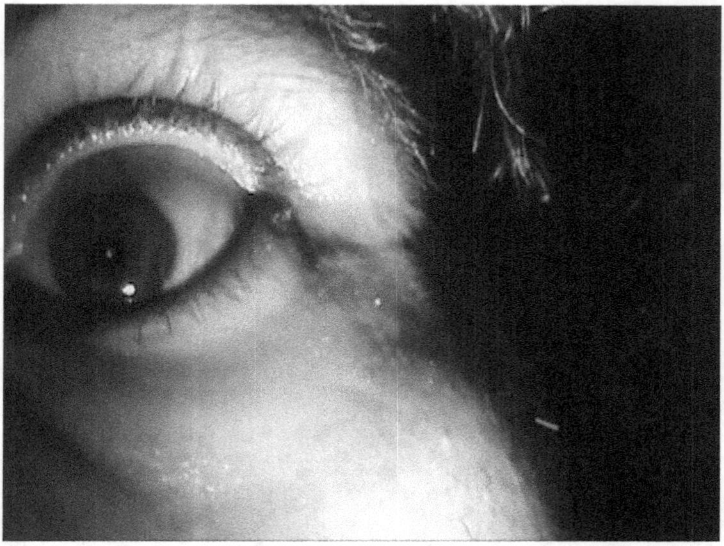

Fig. 5.22 The particle elongates and continues to move in a straight line towards the maxilla.

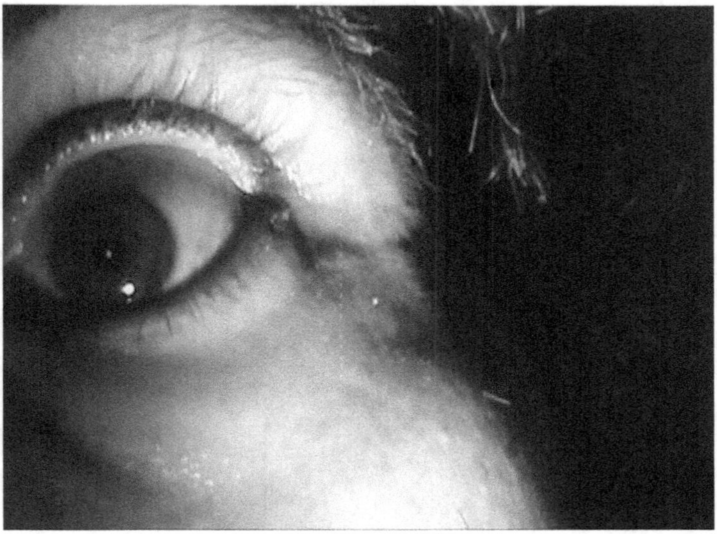

Fig. 5.23 The particle immediately prior to contact with the epidermis elongates to double the length compared to when it first observed. Also note that it is much fainter and that the line of trajectory and the long axis of the particle remain congruent.

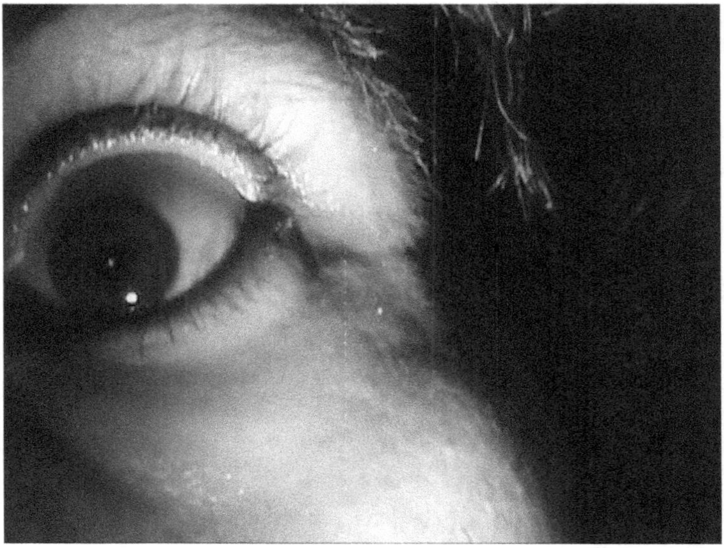

Fig. 5.24 The particle is barely visible less than 33 ms prior to the particle contacting the epidermis.

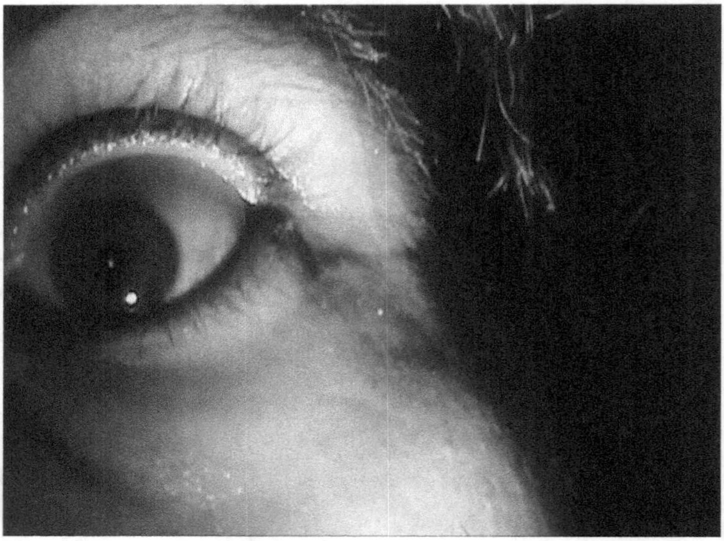

Fig. 5.25 The particle is no longer visible 33 ms after the particle is seen contacting the epidermis.

6

Particles Exiting the Eyes

The ocular phenomenon of white particles exiting the eyes was initially observed in the subject's feline several months prior to recording the videos demonstrating the same phenomenon in the human subject. The only symptom reported by the subject was that of ocular irritation and tearing.

The particles were mainly minute white particles under 1 mm in diameter. The phenomenon was more pronounced after ingesting cocaine. Occasionally, other materials were observed exiting the eyes including a globule of white mucus and a pink substance resembling a cat's claw in shape.

Video 22 This video shows white particles emerging from the eye and then migrating to the eyelashes.

The following six frames are taken consecutively, 33 ms apart. They show an example of particles entering and exiting the eyes.

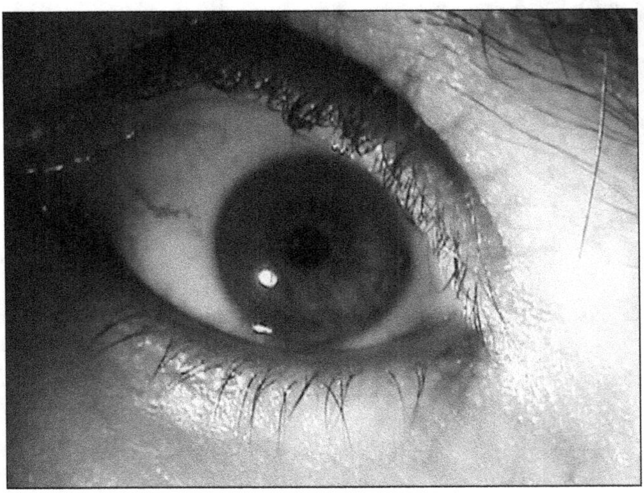

Fig. 6.1 This frame shows several white particles located on the eyelashes and a hair coated with what appears to be many white particles.

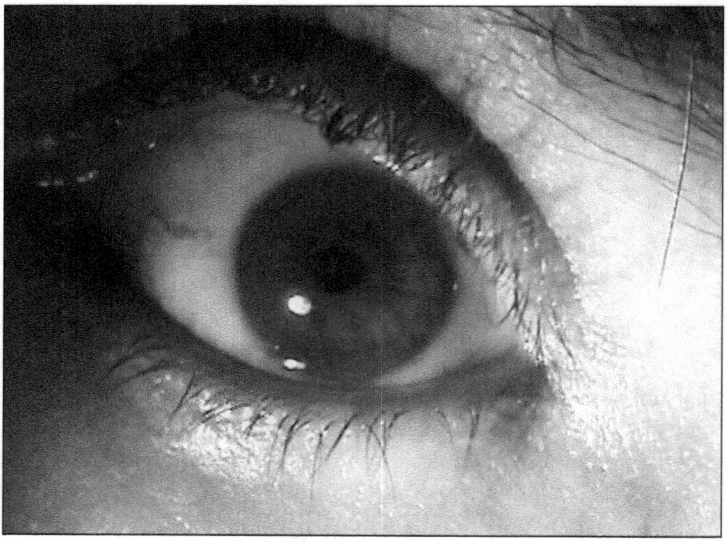

Fig. 6.2 Two particles are seen emerging at the 3 and 7 o'clock positions.

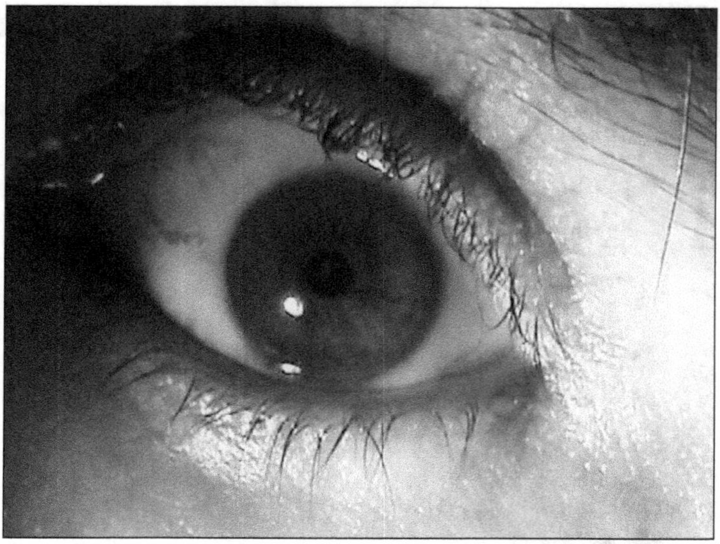

Fig. 6.3 The particles have disappeared 33 ms later.

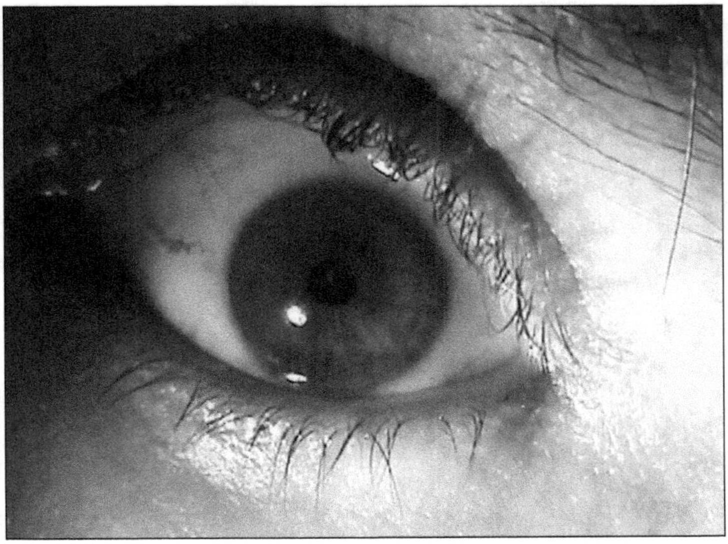

Fig. 6.4 The particle is beginning to reemerge at the 3 o'clock position.

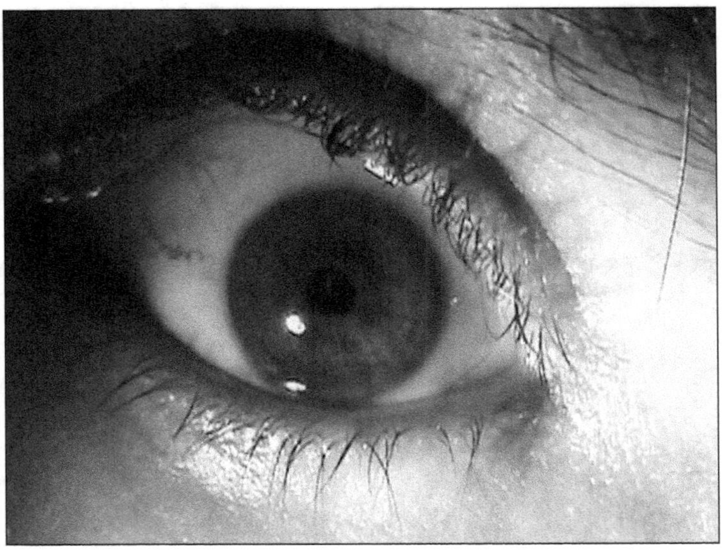

Fig. 6.5 The particle at the 3 o'clock position continues to emerge becoming brighter.

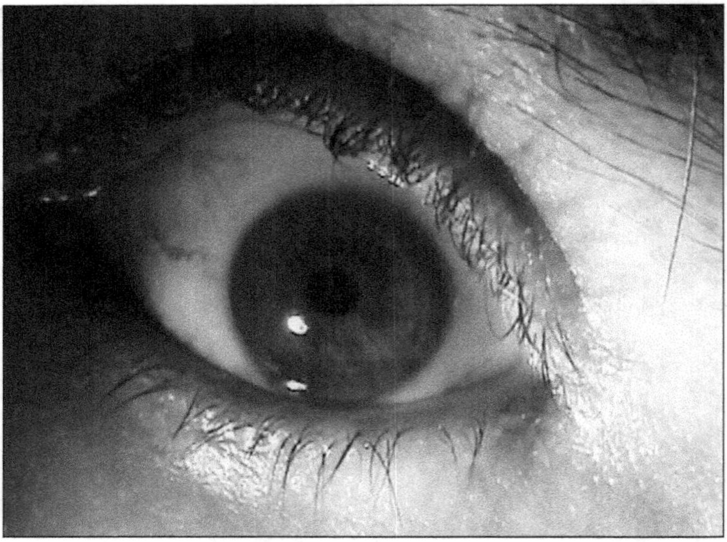

Fig. 6.6 The particle begins to reenter the eye at the 3 o'clock position.

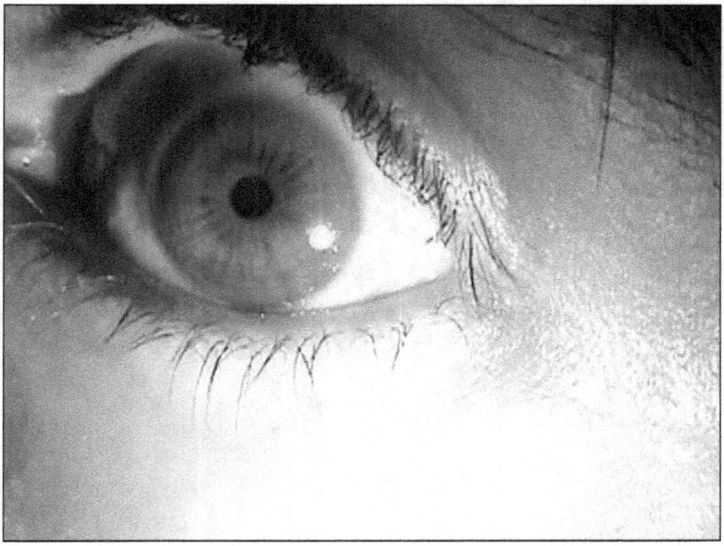

Fig. 6.7 This frame is taken seconds after the sequence above. Over 20 white particles are seen in a patch on the cornea. Video evidence confirms these particles probably emerged from the cornea and sclera. The videos also support the hypothesis that the white particles originate in the eye, emerge, migrate to the inner eyelid margin, cross the lid margin and then migrate onto the eyelashes.

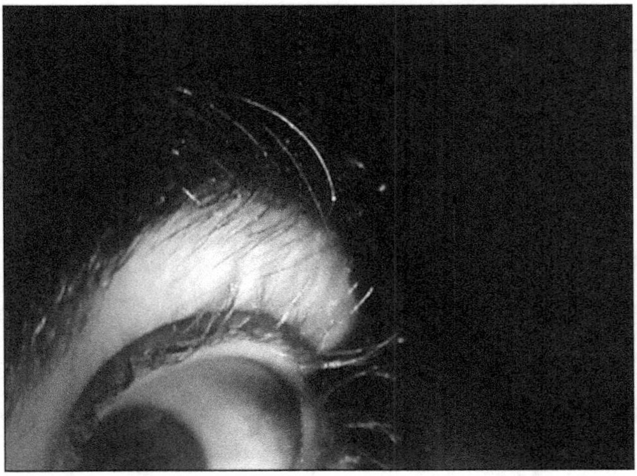

Fig. 6.8 This photograph shows that the eyelashes and scalp hair are coated in white particles.

Video 23 This video shows a white band, several white strands and white particles on the globe.

Video 24 This video shows white particles exiting the middle inferior globe. One of the white particles exits the eye within a tear droplet. It also shows a good view of an eyelash coated in white particles.

Video 25 This video shows white particles exiting the globe at the middle inferior lid margin. Figure 6.9 shows couple of the particles.

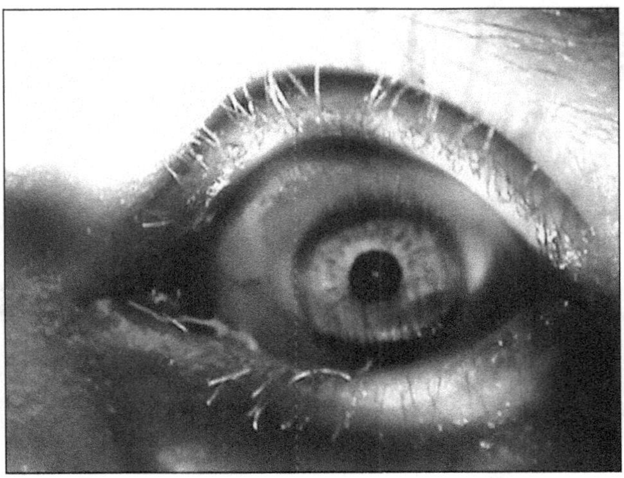

Fig. 6.9 A white strand is seen extending from the lid to the cornea at the six o'clock position. Particles are seen on the cornea over the iris at the 4 and 5 o'clock positions.

A white globule of mucus 2 mm in diameter spontaneously emerges from the inner canthus of the right eye and is immediately placed on a glass slide and examined under a microscope. The results are shown below.

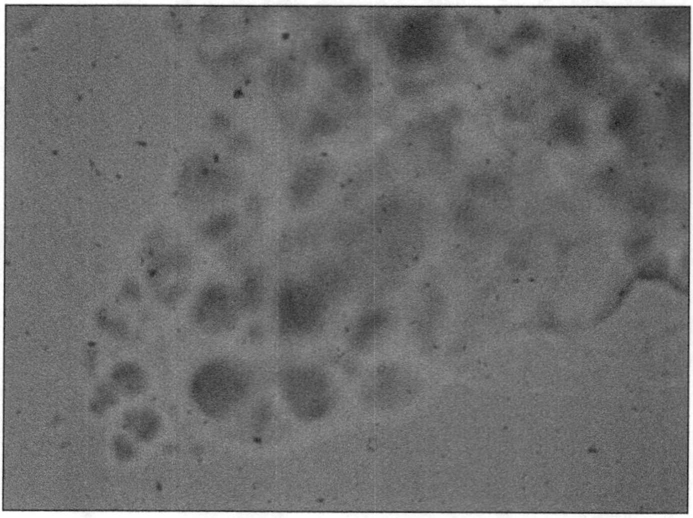

Fig. 6.10 At 400X magnification, the ocular mucus contains multiple smaller globules under normal aperture lighting.

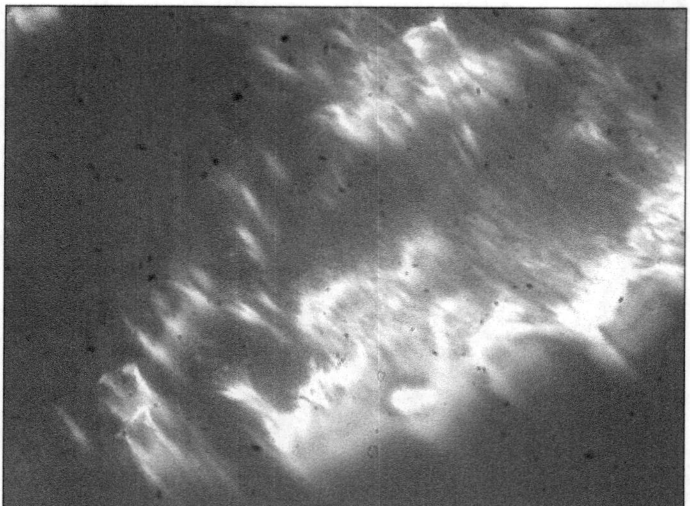

Fig. 6.11 The same view and magnification as figure 6.10 except using the Welch-Allyn 46000A headlight for illumination. The small globules within the larger globule appear to be luminescent.

7

A Substance that Transverses the Optic Globe

Like the particles seen entering the eye, this phenomenon was first discovered while looking for white particles emerging from the eyes. The translucent bands are most likely a continuation of the bands of keratin-microorganism complex that are also traveling through the epidermis of the feet as discussed in chapter 7. The evidence for this assertion is discussed in Chapter 12.

Video 26 This video shows a gap forming between the upper eyelid and the eye globe indicating the existence of an invisible space occupying lesion.

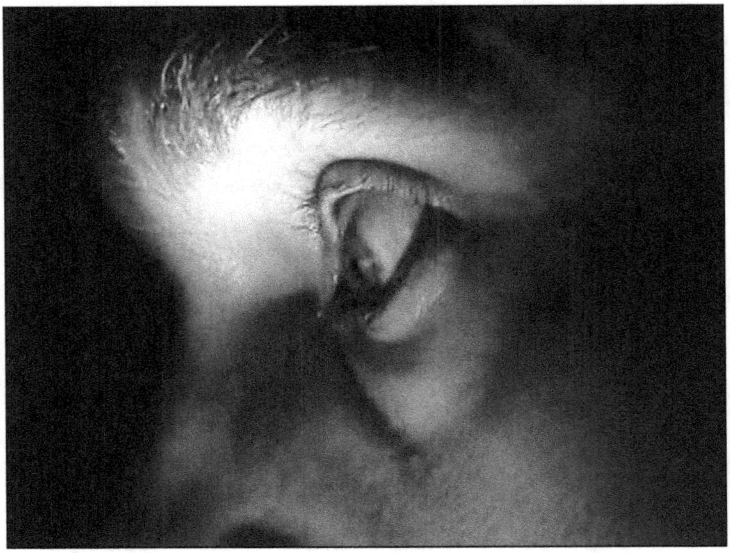

Fig. 7.1 A gap is seen between the upper lid and the optic globe indicative of a space occupying material that is lifting the upper lid from the optic globe. The material is only partially visible under the existent light conditions. A small amount of the material is seen protruding from the gap. Again, as with many of the photographs of the eye, a blue band is seen at the 3 o'clock position of the cornea.

Video 27 This video shows a colored band extending across the eye.

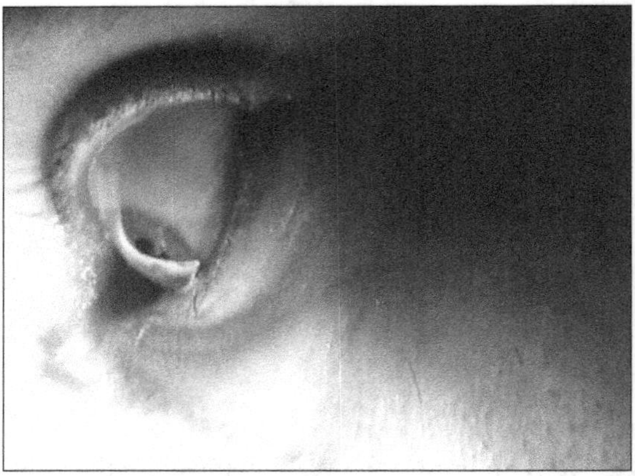

Fig. 7.2 This photograph is an excellent example of the bands that cover the optic globe. Note that the band is seen protruding from the area of the gap in the upper eyelid as seen in Figure 7.1. It extends inferiorly from the gap to the lower lid margin where in becomes thinner. The microorganism-keratin complex contains several pinpoint areas of stronger illumination which most likely represent embedded white particles. The possibility that the substance is mucus is unlikely. Mucus would not have the viscosity or solidity to push the eyelid off the globe.

Video 28 This video shows a translucent band extending across the eye globe from beneath the upper eyelid to the lower lid. The eye lashes are covered in a white material.

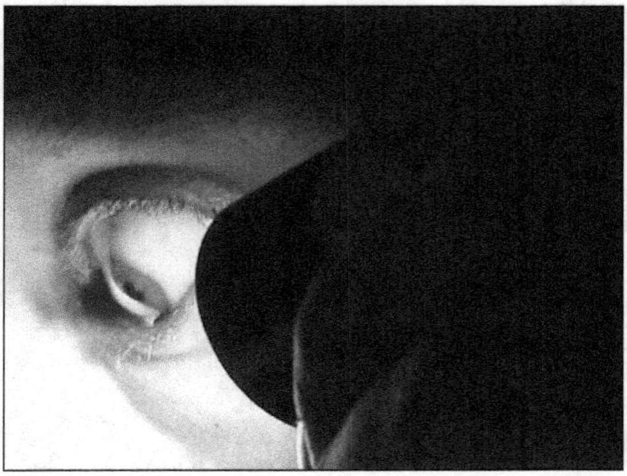

Fig. 7.3 A line rose-colored material is seen only when using a Welch-Allyn 46000A headlight for illumination. The material extends from the mid-upper eyelid to the mid-lower eyelid obscuring the medial aspect of the iris. The superior lateral aspect of the iris has lost its normal curvature and the scleral vessels are not visible, together suggesting the presence of a substance covering the sclera. The material is not visible when the light source is not at this precise angle.

Video 29 This video shows a gap between the globe and the lower lid. A translucent band of material imbedded with white particles extending from the upper lid to the gap in the lower lid is also seen in this video and is only visible when the light from a Welch-Allyn 46000 REVA headlight encounters the band at a specific angle or at specific angles. There are multiple white particles on the globe. The eye lashes are heavily coated with white material and the globe occasionally has a green tinge.

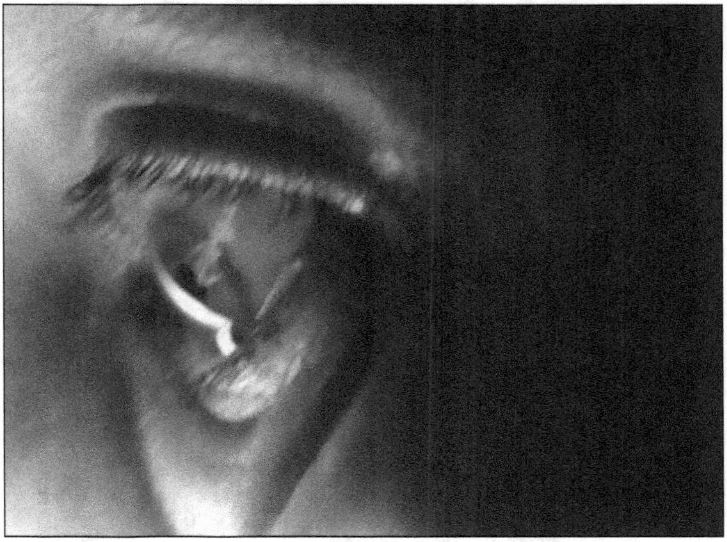

Fig. 7.4 A luminescent and or reflective band of material is seen covering a portion of the pupil with blue coloured bands on the lateral sclera and cornea. There is a blue band extending across the slera and cornea from the upper eyelid to the pupil (see DVD for color photograph). This band is seen In several other photographs of the eye (see Figures 7.1 and 7.2 for example).

Video 30 This video shows two bands extending from the upper lid to the lower lid. The bands are only visible when the Welch-Allyn light source is at a specific angle or angles.

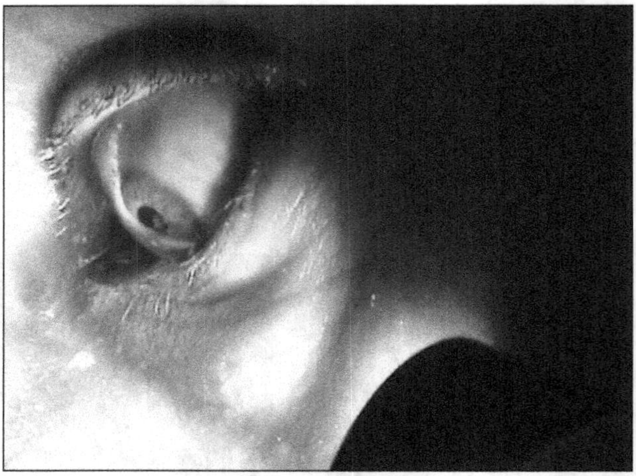

Fig. 7.5 This shows a semi-translucent band with a prominent white particle embedded within it at the 6 o'clock position of the pupil. Also note that the anterior chamber of the eye appears to be filled by the iris and that the iris seems to communicate directly with the cornea as though the material at the 6 o'clock position is coming from the iris. This is the same phenomenon as described in Figure 8.3. The border between the iris and the sclera is concave rather than convex at the 2 o'clock position.

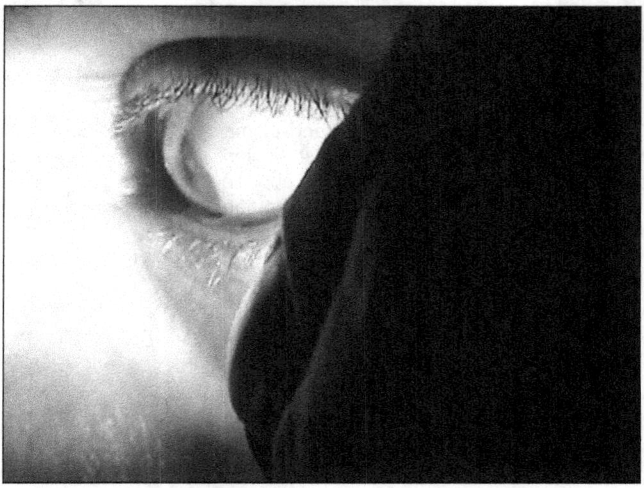

Fig. 7.6 A translucent band is seen extending from the upper lid margin to the lower lid. There are several white particles imbedded in the band. This translucent band of material corresponds to the amphorous material seen in Figures 7.2, 7.3 and 7.5 which also happen to be imbedded with circular particles. There is a green tinged area superior to the iris. The subject's feet were previously covered in ink from a permanent green marker as seen in Figure 4.7 (see DVD for color photograph). There is a concavity of the iridic margin at the 2 o`clock position. The normal convexity of the scleral iridic border is concave at the 2 o'clock position.

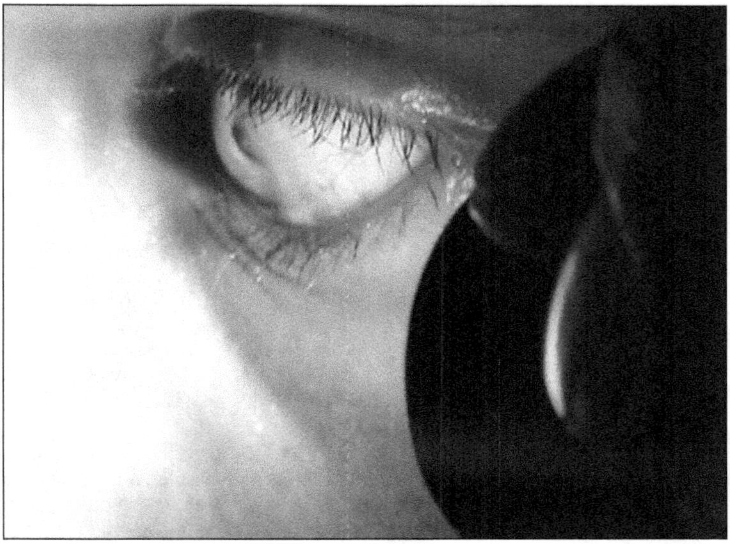

Fig. 7.7 A translucent band extending from the upper lid margin to the lower lid appears to cross over a rose colored band at the inferior lid margin (see DVD for color photograph). Some of the rose colored material is over the lower lid margin. The iris is barely visible between the two bands.

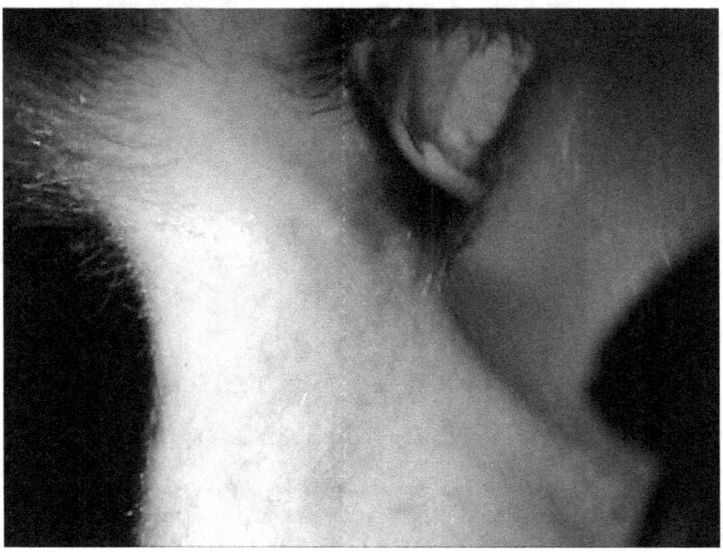

Fig. 7.8 This photograph shows another view of the rose colored band covering the entire iris (see DVD for color photograph). A close examination of the inferior aspect of the sclera reveals that it appears to be covered in a cluster of four circular white plaques.

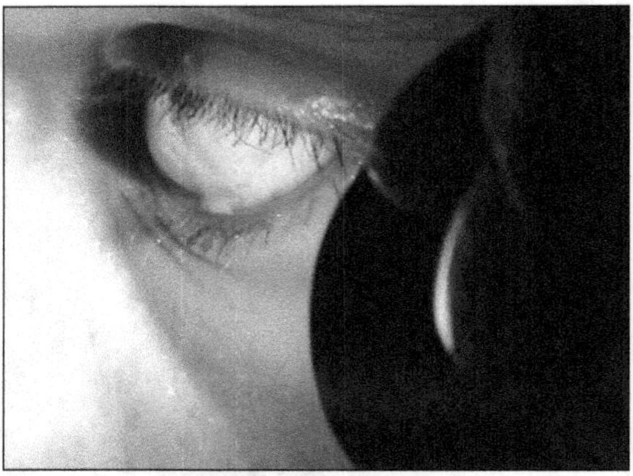

Fig. 7.9 Seconds afterwards, the iris is completely obscured by the rose colored and translucent material. The rose coloration may be from the hemoglobin produced from blood present on the feet analogous to the green color present in the sclera after green ink was applied to the feet. Another source of the pink color of the band may be found in the engorged blood vessels located on the sclera. The arteriole located at the inferior third of the border between the plaque and the sclera is swollen and pink instead of red and thin. This may indicate that the organism is somehow obtaining nutrients from the arteriole (see DVD for color photograph).

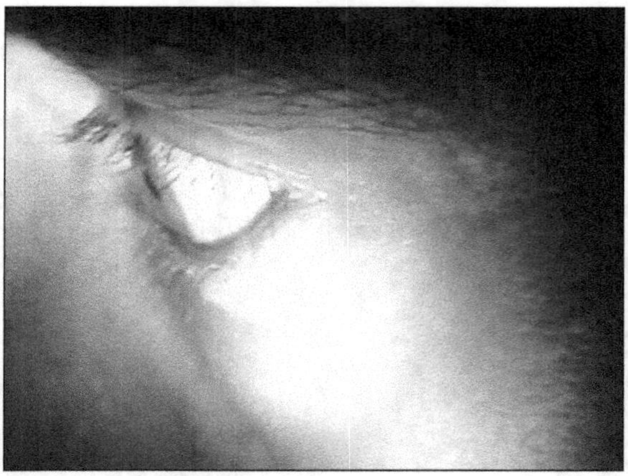

Fig. 7.10 A still frame taken seconds later shows the plaques have completely disappeared. The explanation for this is simple: the beam of light emitted from Welch Allyn headlight is no longer directly illuminating the iris and the beam of light is no longer at the required angle to reflect the light off the micro organism complex, thereby making the plaque invisible. This is one of the reasons why the micro organism has eluded detection until now: it is invisible under almost all lighting conditions. Note that the iris appears to continue to occupy the anterior chamber and is consistent with the iris abutting the cornea. The proximity of the iris to the interior surface of the cornea supports the conjecture that the iris is infiltrated by the microorganism and that it is indeed part of the iris that is seen protruding through the cornea as seen in Figures 8.1 to 8.3, 7.2, 7.5, 7.13 and 7.14.

Video 31 This video shows a translucent band extending across the globe, again visible only with the angled Welch-Allyn light.

Video 32 This video shows a white band covering the medial cornea with a protruding papule. Note that the globe occasionally appears to have a greenish tinge and that white particles are exiting the globe.

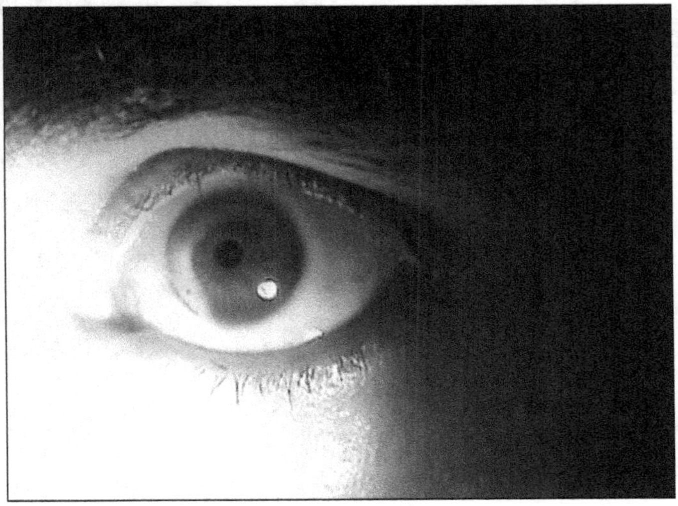

Fig. 7.11 A white band with a dark area at the 8 o'clock position is seen on the medial cornea.

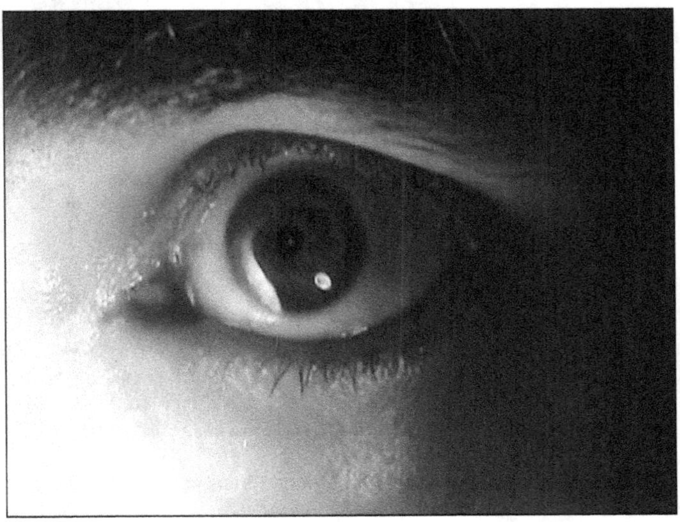

Fig. 7.12 The same white band however the dark area now appears to be a protruding white papule.

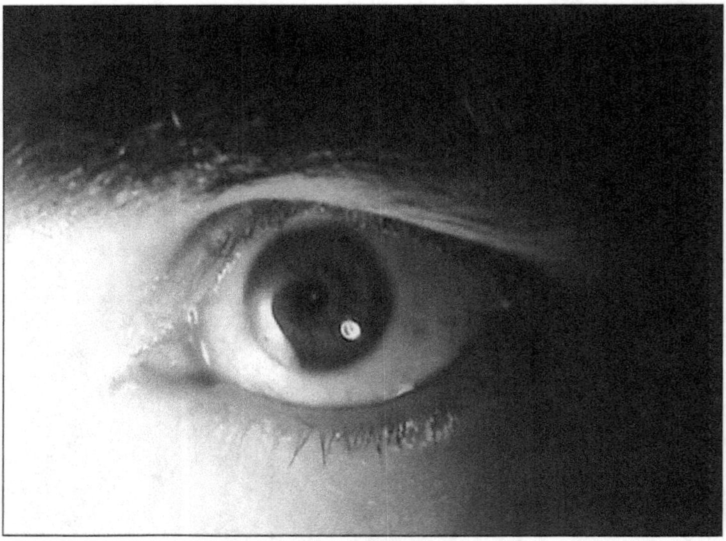

Fig. 7.13 The same white band with the protuding area now appearing to be partially dark. There also appears to be several luminescing white particles at the upper medial lid margin and the lower lid margin lateral to the midline.

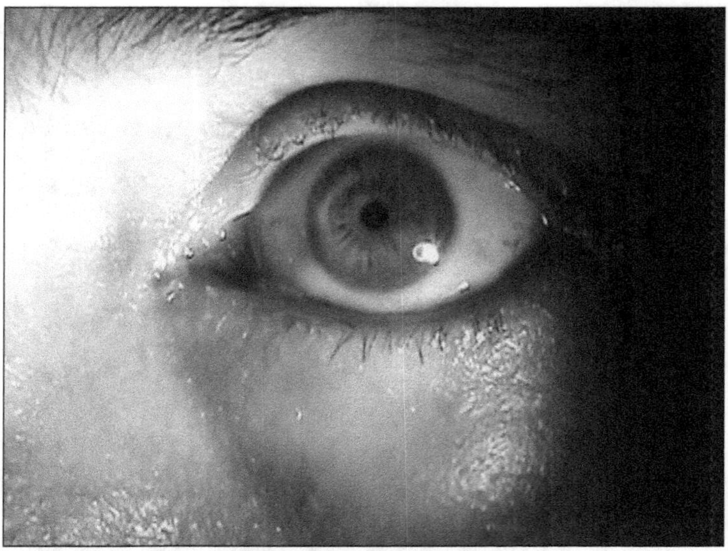

Fig. 7.14 The same white band with a green tinged area at the 4 and 5 o'clock positions that appears to have lost the normal scleral appearance.

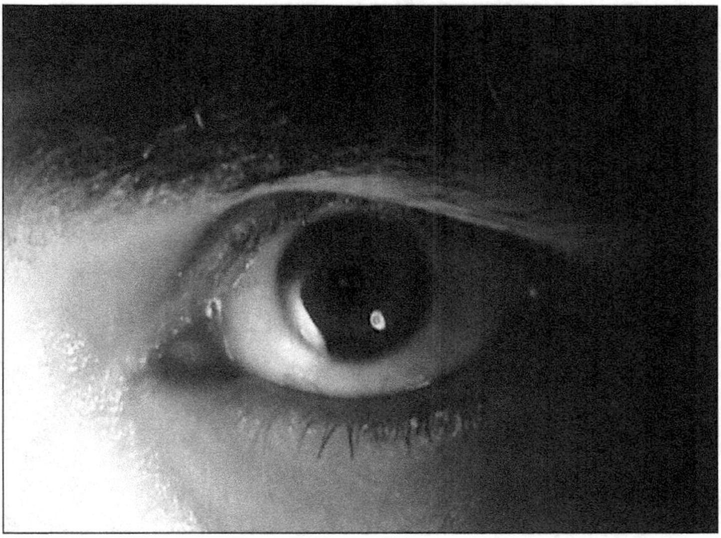

Fig. 7.15 The same as Figure 7.14 with what appears to be several more protruding papules an the sclera at the seven o'clock position.

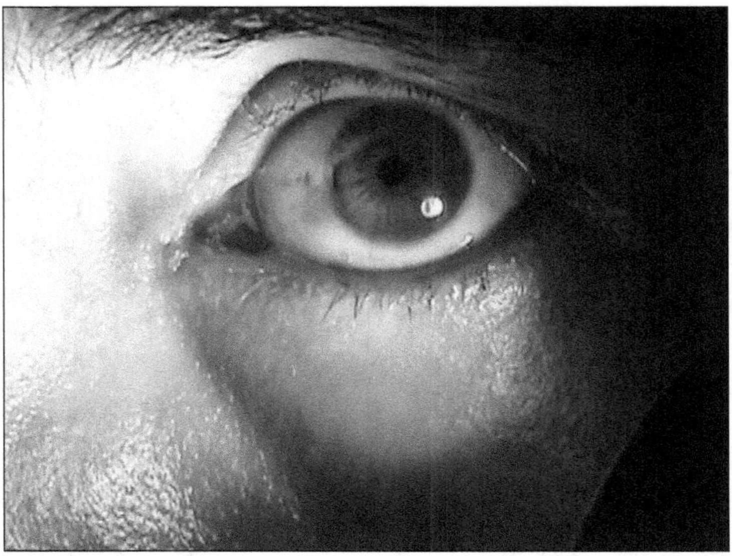

Fig. 7.16 This frame shows a black area on the medial sclera.

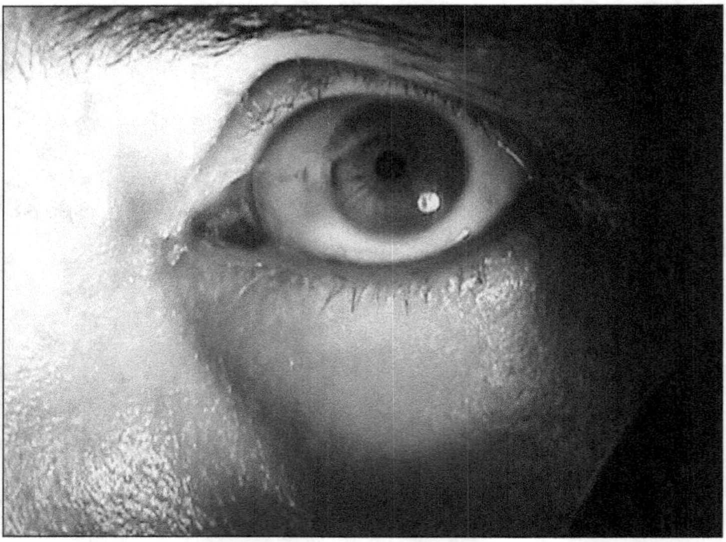

Fig. 7.17 There are white materials at the 8 o'clock position and a white band can be seen on the medial sclera. Several minute white particles can be seen on the sclera medial to the iris.

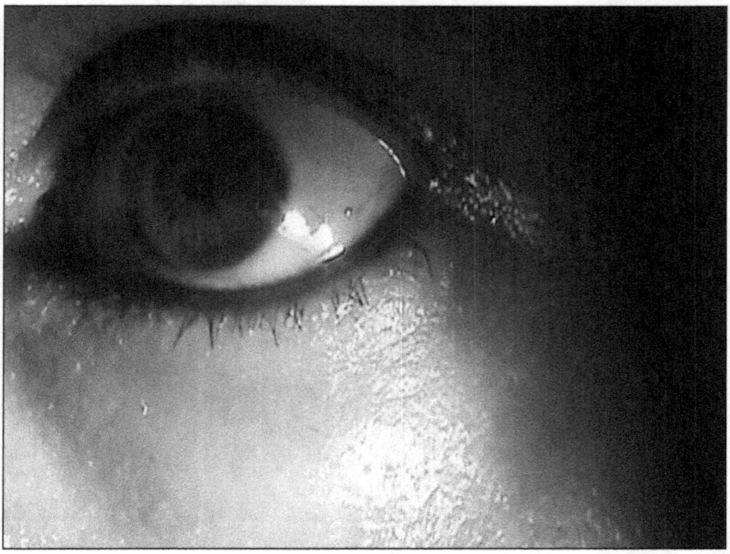

Fig. 7.18 A luminescing particle surrounded by black is seen on the lateral sclera.

8

The Irregular Iris

The changes in the iris of the subject's feline were also observed in the human subject. Although a similar phenomenon involving the iris was described earlier in Chapter 7, the phenomenon elicited in this video is considered an important enough phenomenon to warrant a separate chapter.

The following video shows a more compelling view of the role of the iris in the generation of particles. It can be clearly seen that the iris is abutting the cornea and emitting particles into the bands covering the cornea.

Video 33 This video shows black particle near the pupil and a white particle connected to another black particle located on the inferior edge of a translucent band extending across the cornea. The appearance of these particles was associated with severe eye irritation.

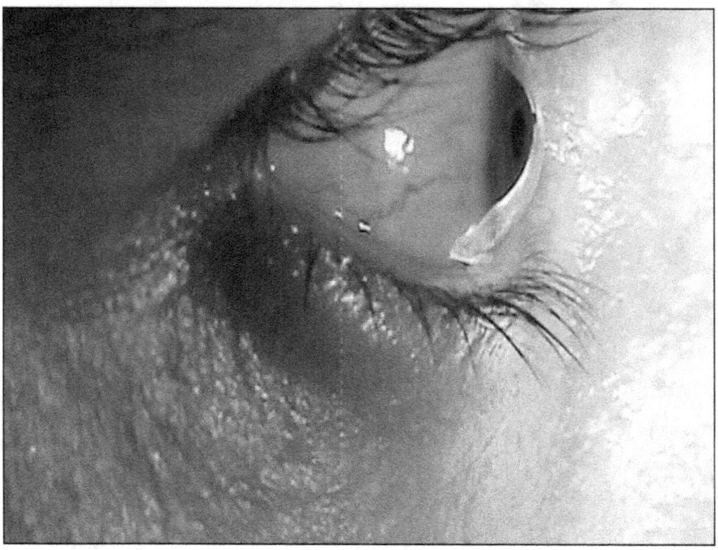

Fig. 8.1 This photograph shows the phenomena described in the video description above.

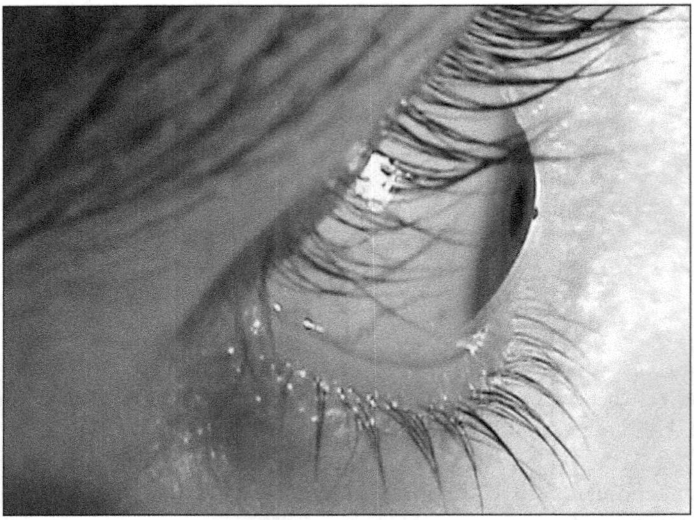

Fig. 8.2 This frame shows several interesting phenomena. A black material is seen emerging from the cornea. The superior lateral curvature of the iris is concave rather than convex. There is a luminescent substance at the inferior mid lower eyelid margin. There is a green substance in the lower eyelashes that probably originated in the epithelium of the feet which were previously covered in a green ink (see Chapter 12).

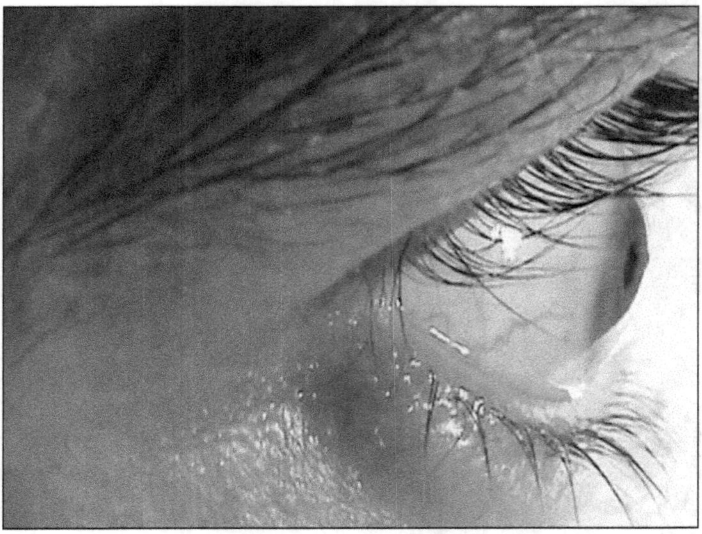

Fig. 8.3 Another view of the dark area suggests that it is contiguous with, and originates in, the iris. Note that the pupillae sphincter muscle margin is irregular and the iris fills the anterior chamber, appearing as though the entire iris is being pulled into the protrusion. Also note that the sphincter pupillae muscle (located at the interior border of the iris) is indented at the same location as the protrusion.

9

Particles that Exit the Feet

The phenomenon of particles exiting the feet was the most common phenomenon observed as well as the most difficult to argue has actually occurred. The reason it is difficult to argue is that it is easy to dismiss the video evidence as dust particles blowing in the breeze. Despite this, it is hoped that the evidence presented here will demonstrate that particles are exiting the feet in the presence of vaporized cocaine.

Video 34 This video shows particles exiting the feet. A frame by frame analysis shows 2 particles exiting the lateral distal plantar surface. The following figures are taken 33 ms apart and show the particle originates from within the dermis and changes direction once it is airborne.

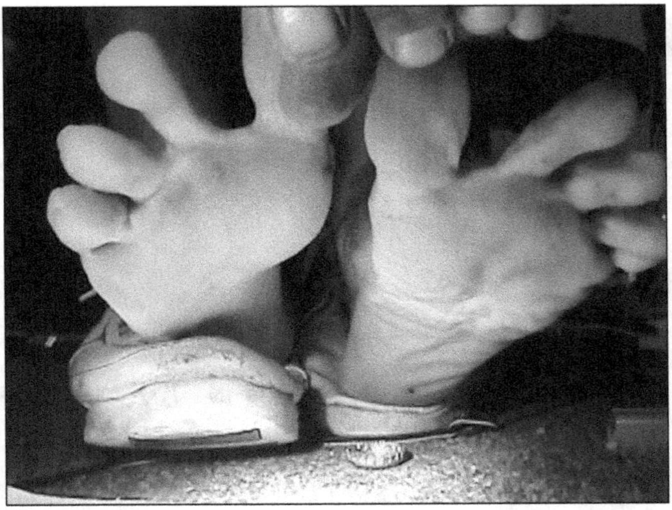

Fig. 9.1 This photograph shows a view of the foot 33 ms prior to the appearance of the subject particle. The foot has just been exposed to vaporized cocaine. Another particle is seen traveling through the air on the left of the photograph. This particle appears to contain two parallel sections that move and change direction as a unit. At the bottom far left of the photograph, a piece of paper can be seen and it does not move throughout the sequence of photographs indicating that the particles are not propelled by air currents.

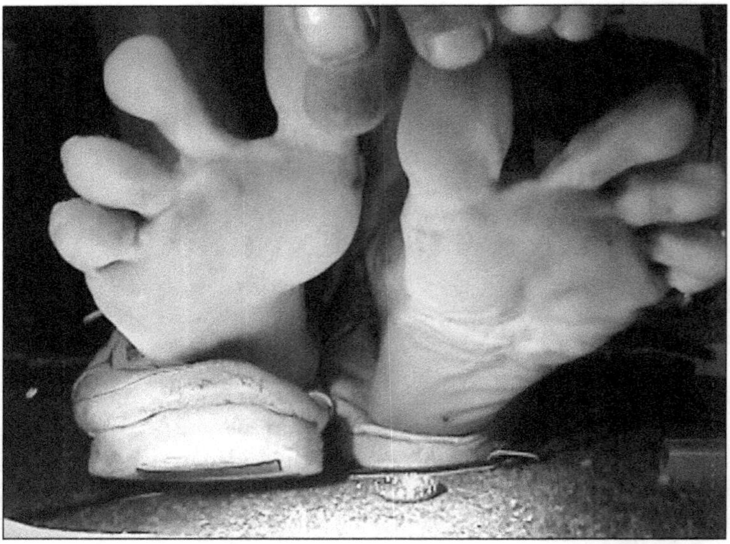

Fig. 9.2 This photograph shows the first appearance of another white particle on the right foot proximal to the fourth and fifth phalanges. Note that the particle seen in Figure 9.1 continues to travel through the air towards the 300 watt source of halogen illumination with congruency of its line of trajectory and its long axis.

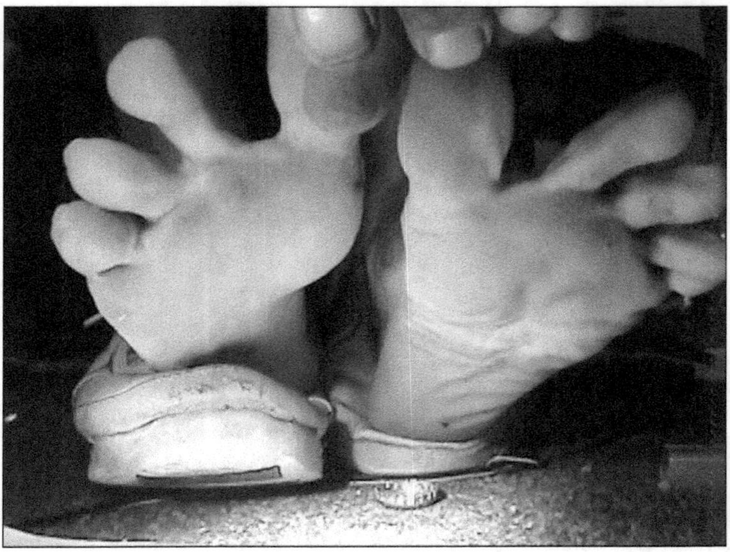

Fig. 9.3 This photograph shows the two particles 33 ms after Figure 9.2. The first particle continues to travel towards the light source with its line of trajectory and long axis congruent, making wind as the source of propulsion unlikely. The second particle is more visible and is moving in the same direction as the first particle, towards the halogen light source.

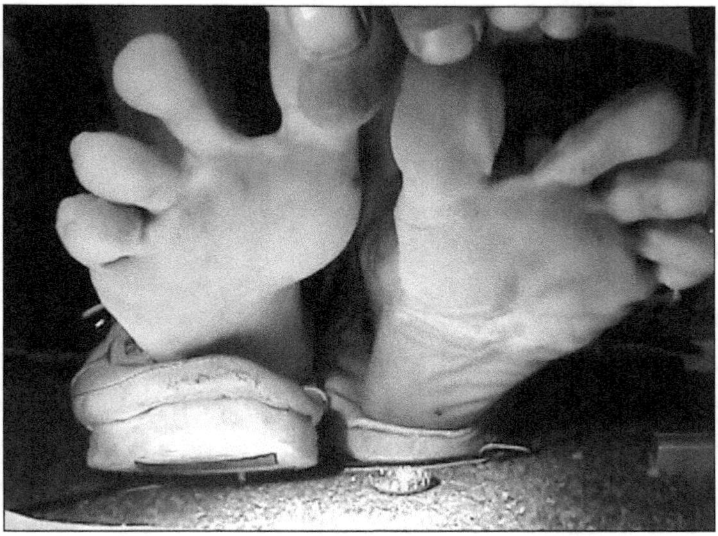

Fig. 9.4 This photograph shows the particle leaving the foot and moving at a 35 degree angle relative to the floor. In almost all of the videos of white materials spontaneously exiting the feet after exposure to vaporized cocaine, the white materials have the same appearance and orientation. The typical appearance of these particles is that of a straight linear mass with the longitudinal axis of the particle aligned with its line of trajectory.

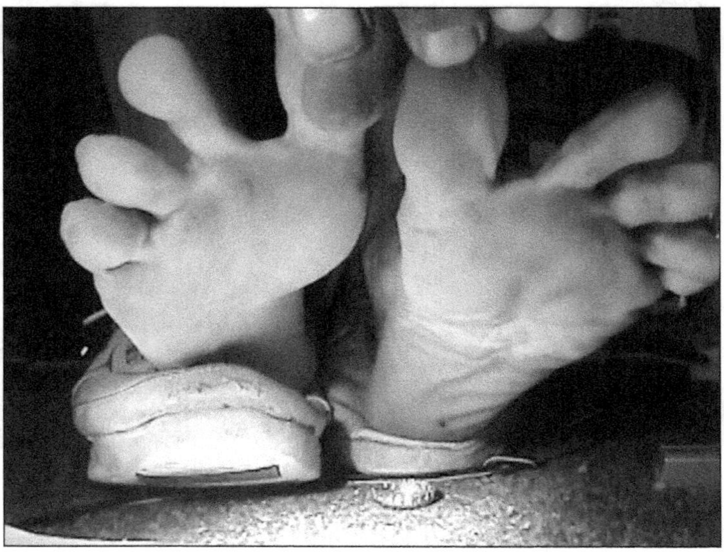

Fig. 9.5 The trajectory of the particle has changed and the long axis of the particle has also changed accordingly to approximately 60 degrees relative to the floor.

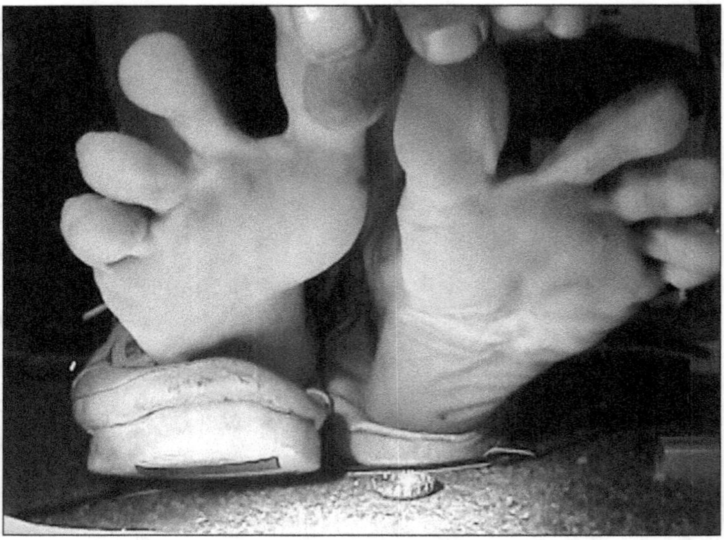

Fig. 9.6 The long axis of the particle continues to be congruent with the line trajectory.

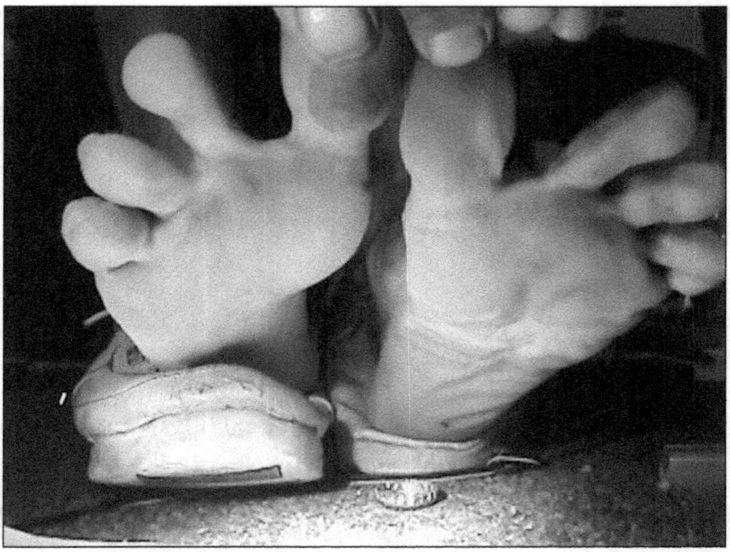

Fig. 9.7 The line of trajectory changes but remains congruent with the long axis of the particle.

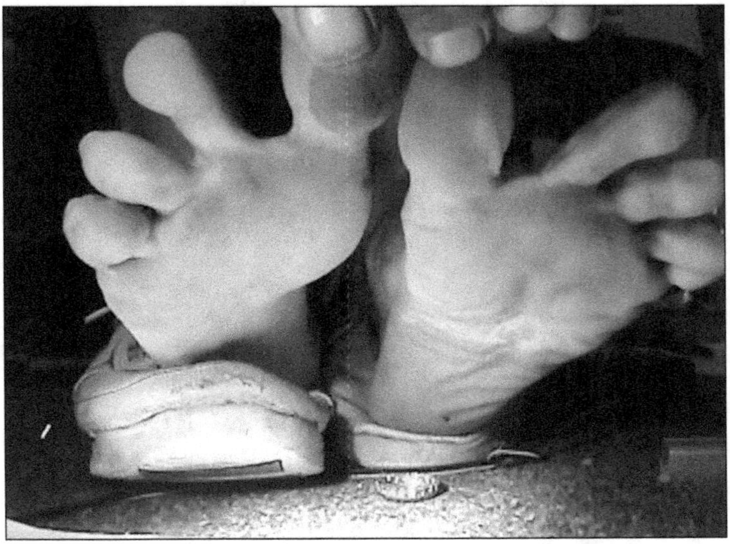

Fig. 9.8 The particle begins to change direction and lengthens temporarily. The long axis of the particle is again congruent to the line of trajectory.

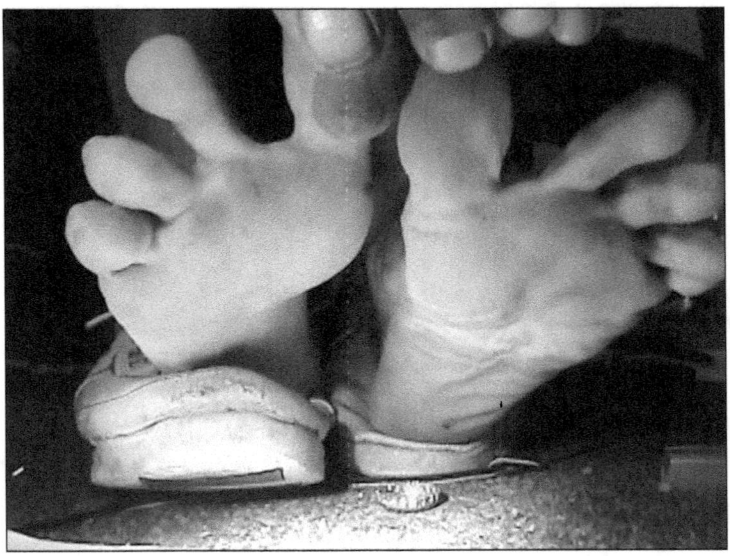

Fig. 9.9 The particle continues to change direction and the line of trajectory and the long axis of the particle are still congruent.

Video 35 This video shows a white substance being extracted from the foot with tweezers. The extracted material from this and similar extractions has a rubbery consistency when it is first extracted. Within 10 minutes of being outside the body the material becomes hard and non-malleable, like thick dried skin. Microscopic examination of these particles is seen in Figures 9.12 and 9.13.

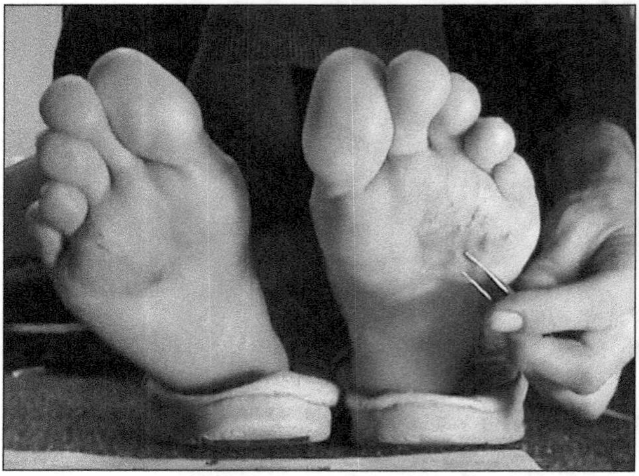

Fig. 9.10 A papule of white material as it appears prior to its extraction with forceps.

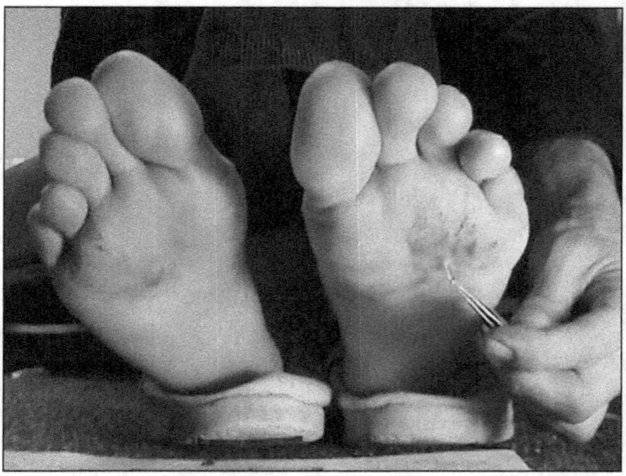

Fig. 9.11 The material has a rubbery consistency. Note that the feet are swollen over the metatarsal heads and the toes appear to be squeezed together. This supports the hypothesis that the micro-organism travels along the plantar surface in a distal direction towards the toes, travels through the toenails and then begins to move proximally towards the head. When the toes are squeezed together the feet swell temporarily for the duration of the contraction probably because the micro-organism's route is temporarily blocked by this action. Within minutes of relaxing the toes the foot swelling is eliminated.

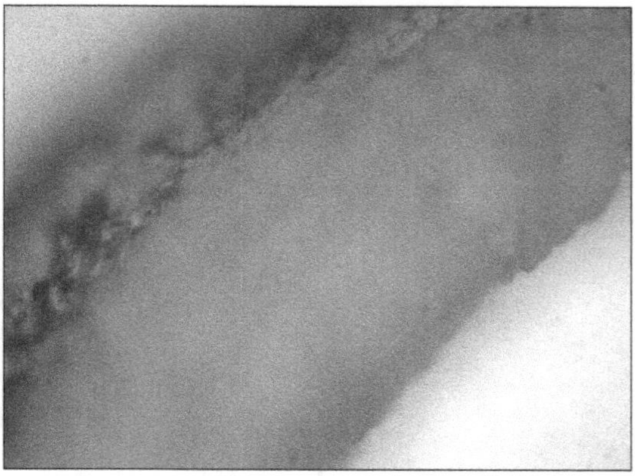

Fig. 9.12 A photograph of a section of material extracted from the foot shows keratinized epithelium. Within the amphorous material located between the stratum corneum of keratinized epithelium on the right and the stratum lucidum on the right are numerous minute circular objects that could be the spores of the organism. These circular objects could be the same circular objects seen in the amphorous material seen on the sclera in Chapter 7. They could also be melanocytes, red blood cells or some other material. There is an irregular band of black material between and parallel to the surface of the epithelium and the stratum lucidum.

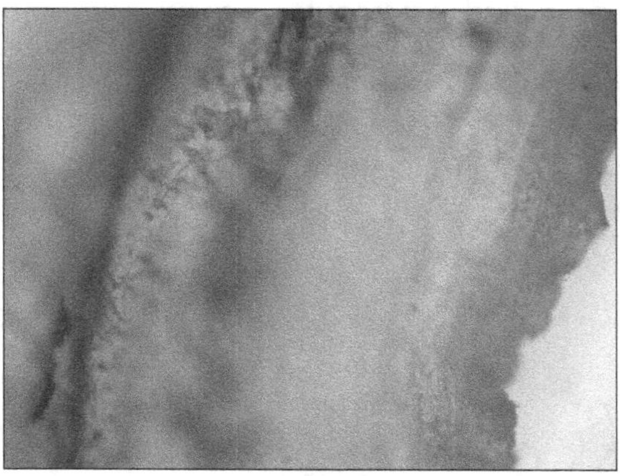

Fig. 9.13 A photograph of the same material as seen in Figure 9.12 as viewed under a different focus at a different location. This view shows that the material removed from the foot is actually a section of dermis. The stratum corneum is seen on the right side and the stratum lucidum is seen on the left. Between the strata is an amphorous area with a dark material within it. Normally, the epidermis does not contain an area of black material between the surface and the stratum lucidum. Also of possible significance is the presence of several minute circular structures located within the amphorous material. There is also a section of dark filamentous material seen on the bottom far left of the photograph. Either of the two areas containing dark filamentous material in this photograph – the one to the left of the stratum lucidum as well as the one traversing the entire photograph within the amphorous material - may correspond to the dark filamentous material seen in Figures 13.4, 13.6, 13.9 and 13.11.

Video 36 This video shows white particles exiting the feet.

Video 37 This video shows a hair, possibly from a tuxedo cat, exiting the foot. The hair follicle is possibly from a cat thought to be infected by the microorganism. The cat was handled by the subject approximately 30 minutes earlier.

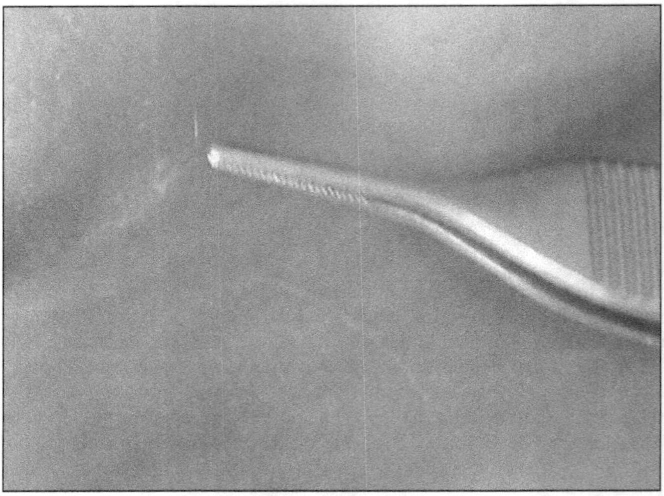

Fig. 9.14 A hair is seen attached at one end to the plantar surface of a foot. The next three photographs are taken over the next three seconds.

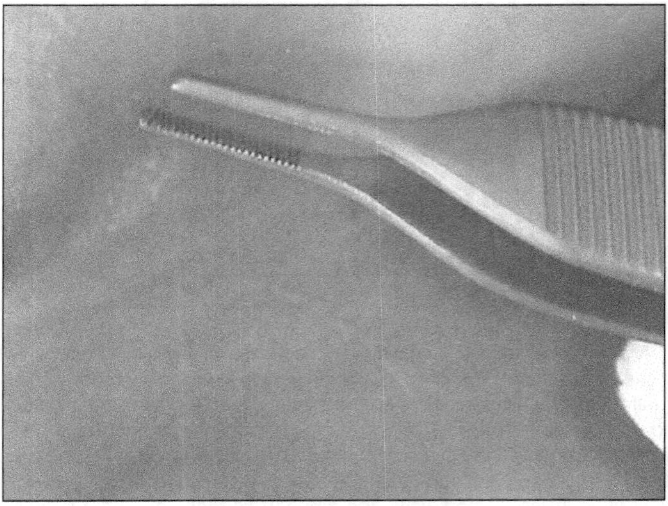

Fig. 9.15 An attempt at removing the hair with a pair of forceps fails. The hair is at a 45 degree angle to the plantar surface and if it would be at a 0 degree angle it would be just a fraction beyond the white particle located above and just lateral to it. Therefore the length is from the origin to just beyond the white particle.

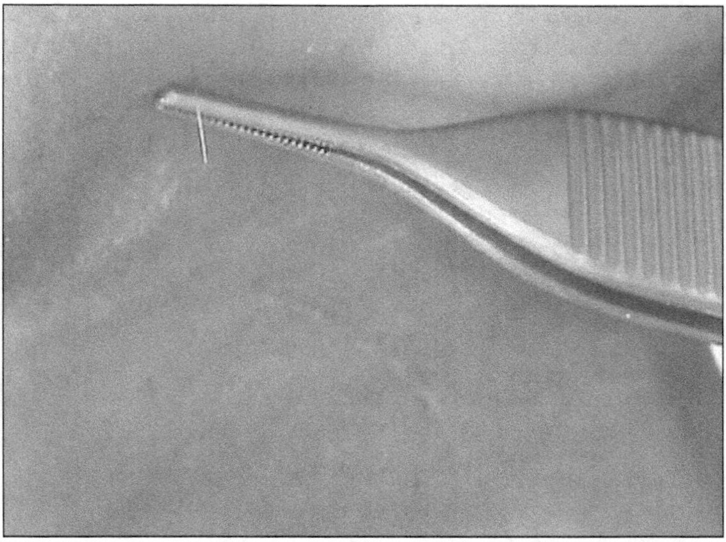

Fig. 9.16 When the hair returns to its original position it is about 50% longer than it was in figure 9.14 using the distance between the origin and just beyond the white particle as the original length of the hair.

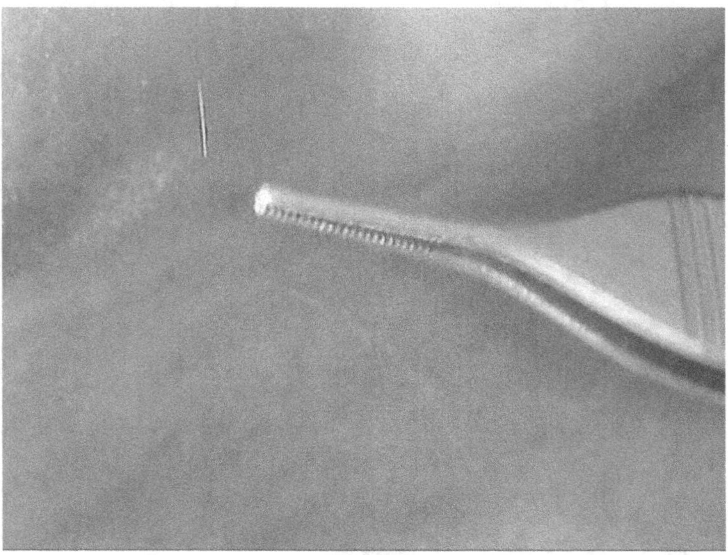

Fig. 9.17 This photograph shows an unobstructed view of the hair.

Video 38 This video shows long white strands exiting the feet with minimal stimulation with tweezers and a particle being removed from the foot that becomes a long strand. Frame by frame analysis of the latter example is included.

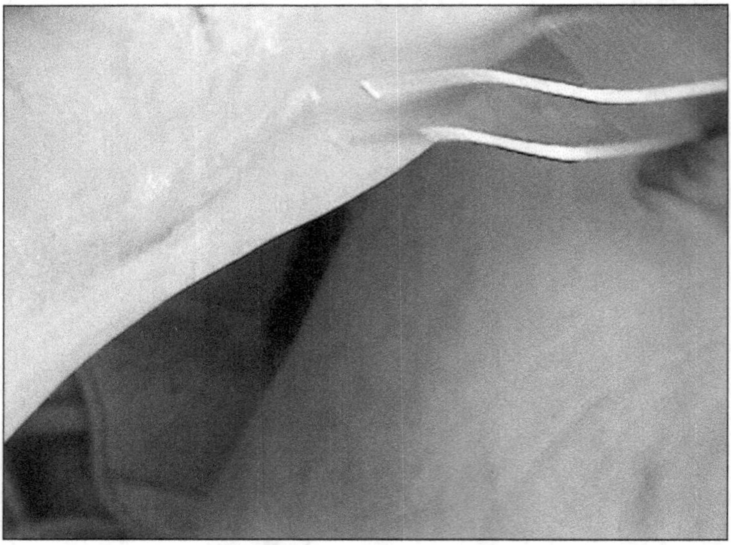

Fig. 9.18 A white particle is about to be removed from the plantar surface of the foot.

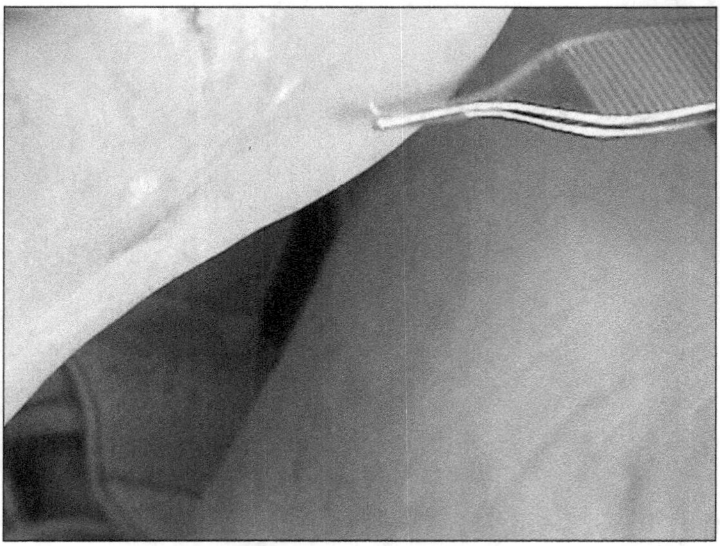

Fig. 9.19 This photograph shows a white particle in the forceps that was just removed from the plantar surface of the foot.

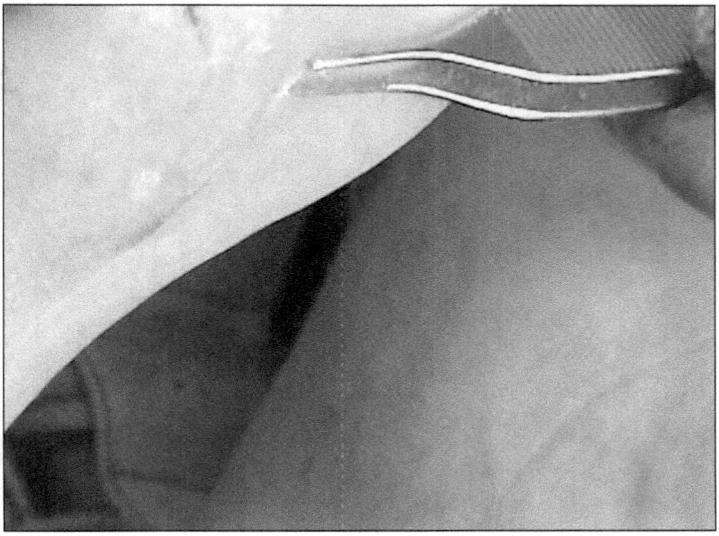

Fig. 9.20 The white particle is returned to the foot surface just superior to the spot from where it was extracted. The next 3 frames are taken 33 ms apart.

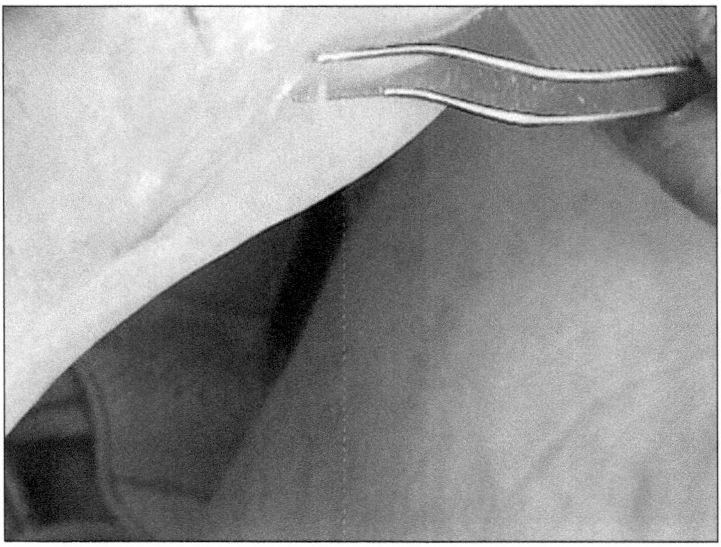

Fig. 9.21 The particle is moving towards the floor.

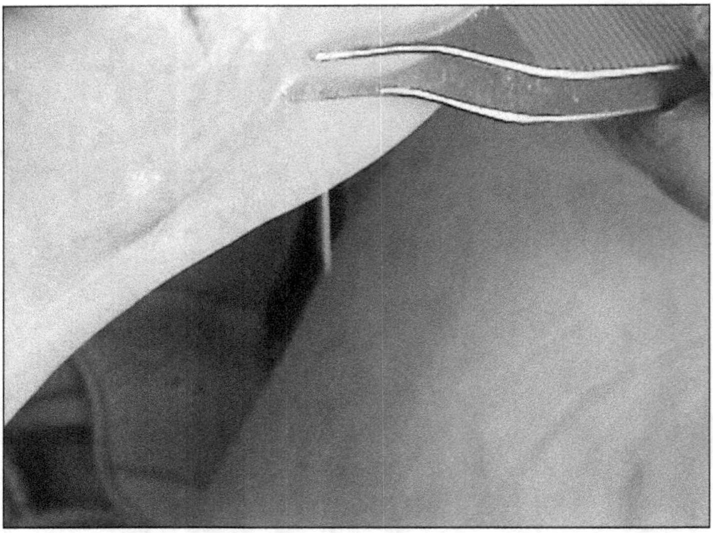

Fig. 9.22 The particle has elongated into a strand.

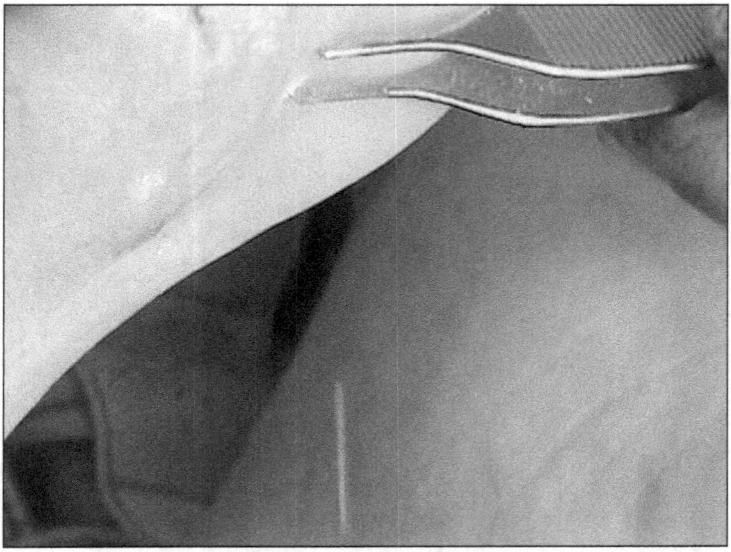

Fig. 9.23 The particle continues to elongate and descend as a strand.

Video 39 This video shows particles exiting the feet.

Video 40 This video shows particles exiting the feet after being stimulated with wind.

Video 41 This video shows particles exiting the feet.

After it was discovered that the coated hair follicles from the forearm (chapter 13) were associated with luminescent white material under microscopic examination, the possibility that the previously collected white material from the feet may also be luminescent was raised. Subsequently, several samples of white material that was collected on sheets of paper placed under the feet months earlier were examined microscopically using the Welch-Allyn light source. As anticipated these particles were strongly luminescent (see Figures 9.24, 9.27).

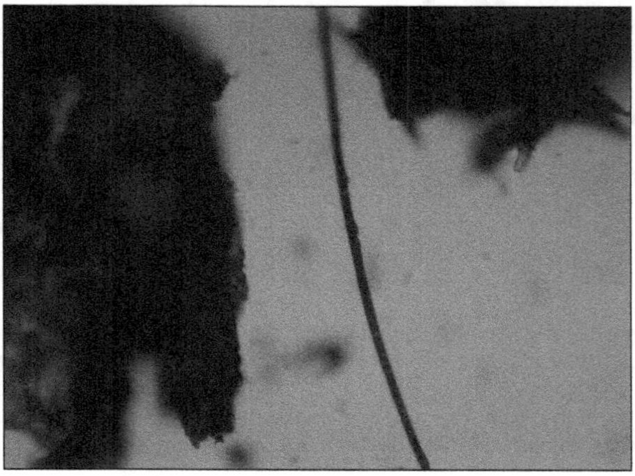

Fig. 9.24 Specimens collected from materials extracted from the plantar surfaces of the feet as viewed under 100X magnification.

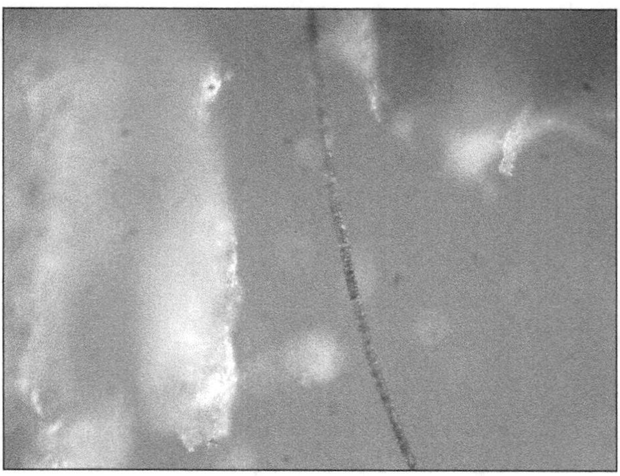

Fig. 9.25 The same material as seen in Figure 9.24 illuminated using the Welch-Allyn 46000 REVA headlamp. There are large areas of strong luminescence indicating that the materials extracted from the feet are probably the same organism that is attached to the hair follicles.

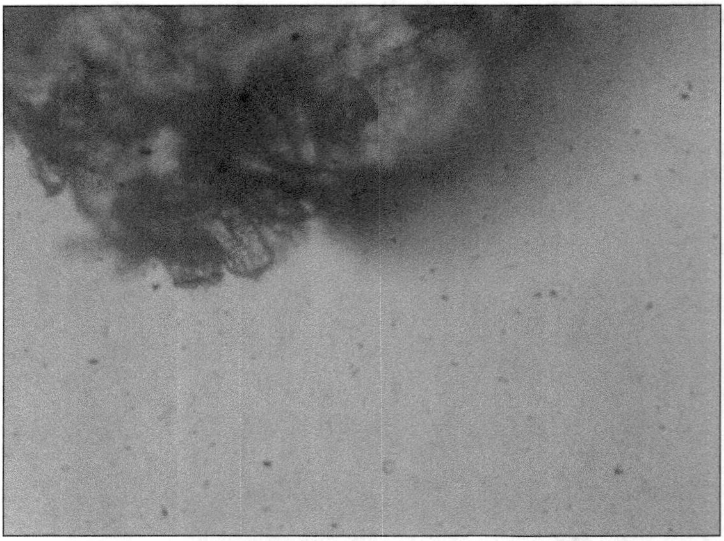

Fig. 9.26 Some of the material extracted from a foot under 200X magnification demonstrates that these materials are composed of an intricate matrix of intertwining fibers under normal illumination. The fibrous structure of the microorganism accounts for the elongation seen while it travels through the air. The furled fibers making up the lace like matrix probably unfurl making the microorganism longer.

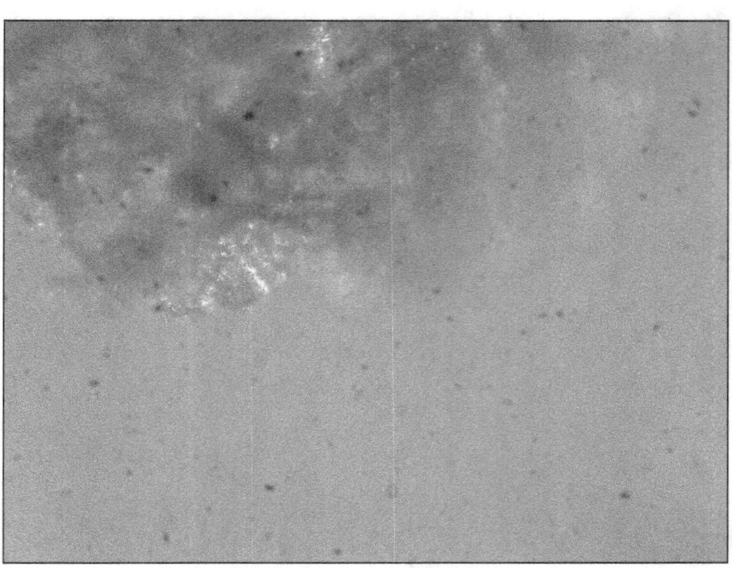

Fig. 9.27 The same material as seen in Figure 9.26 illuminated using a Welch-Allyn 46000 REVA headlight shows areas of luminescence.

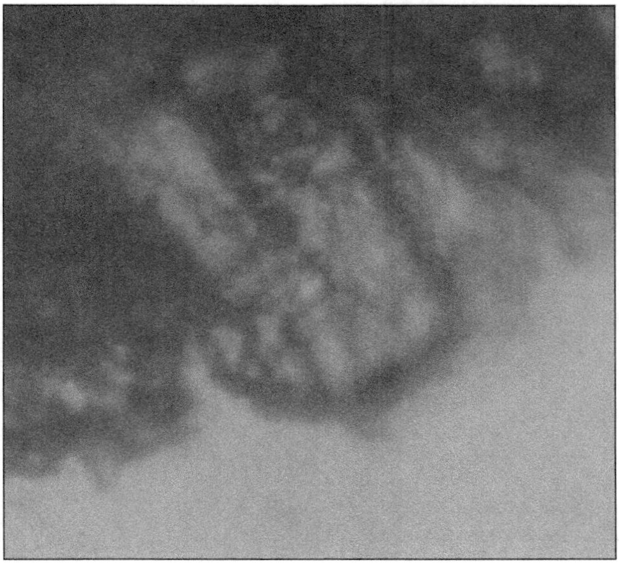

Fig. 9.28 A closer view of the material seen in Figure 9.26 demonstrates the intricate lattice contained in the material extracted from the feet.

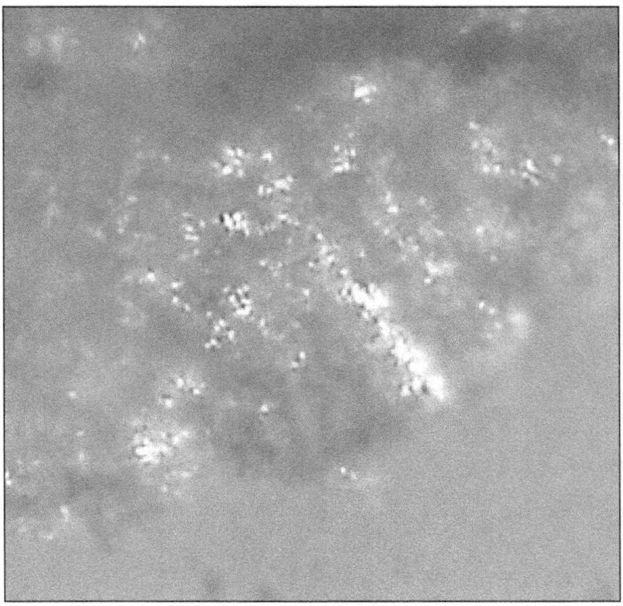

Fig. 9.29 The same view as Figure 9.28 illuminated by a Welch-Allyn 46000 REVA headlight. There are areas of strong luminescence located along the black filaments.

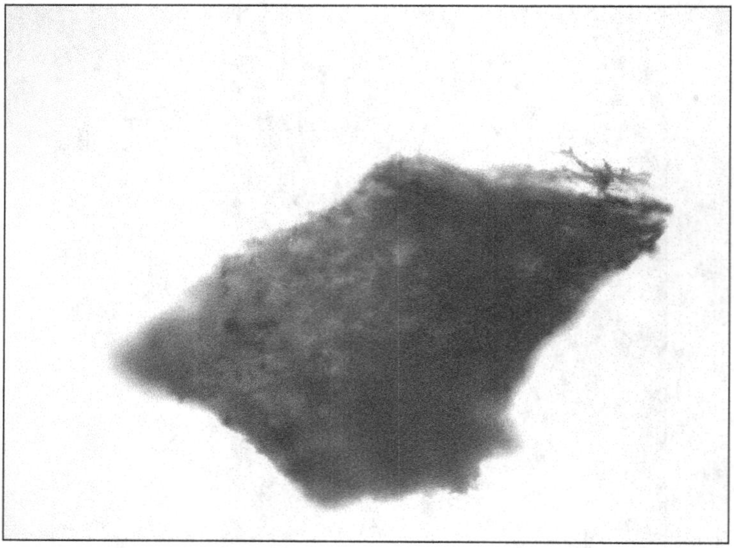

Fig. 9.30 This a photograph of a particle that came out of the foot after a stimulus of cocaine vapours was applied at 100X magnification. This may represent the so-called "coca mite."

10

Particles that Enter the Feet

Other than hair entering the feet (see Chapter 4), several other materials were videotaped entering the feet. White materials initially extracted from the feet have been videotaped reentering the feet at a different location. Avian feathers and lint have also been presented here as particles that have reentered the feet indicating that the micro organism probably attaches itself to non-host materials as well.

Video 42 This video shows white material being extracted from one part of the foot and entering another part. A frame by frame analysis shows that the particle disappears in 33 ms.

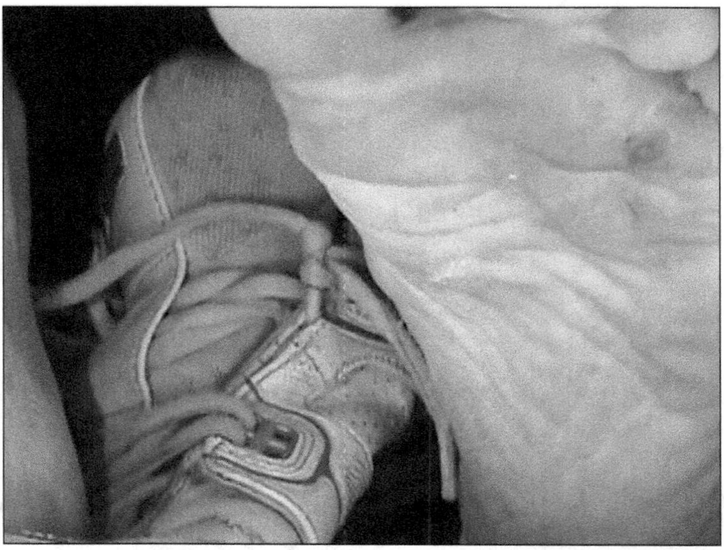

Fig. 10.1 This photograph shows the white particle as seen prior to its absorption into the dermis. Note that the dermis of the plantar surface of the foot has been compromised and blood has conjugated on the epidermis. This may explain the rose color of the bands seen in the eyes in Figures 7.7 to 7.9 (see DVD for color photograph).

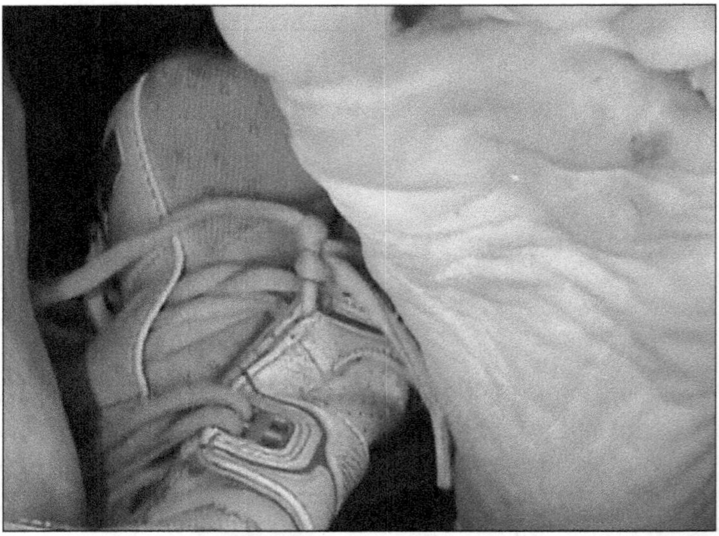

Fig. 10.2 A photograph taken 33 ms later shows a black area appearing superior to the white particle.

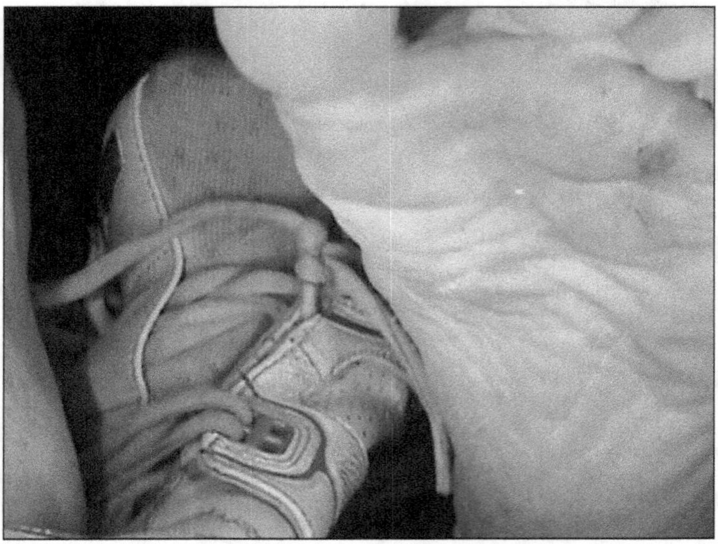

Fig. 10.3 The white particle as it appears immediately prior to entering the dermis.

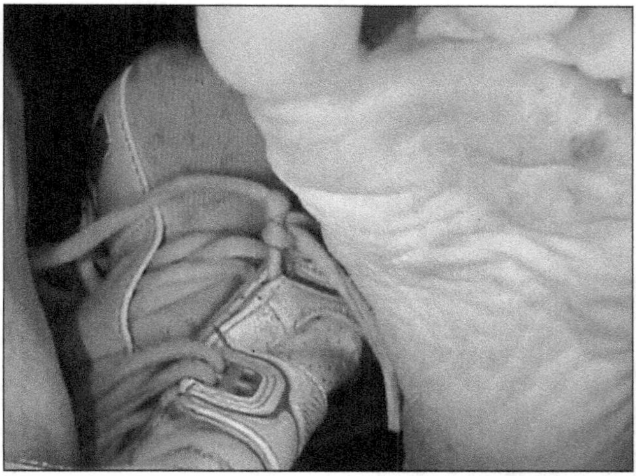

Fig. 10.4 The same view 33 ms after Figure 10.3 shows that the particle has been absorbed into the dermis. The particle could not have fallen or it would still be in the field of view as was seen with the falling particles seen in Figures 9.20 to 9.23 and Figures 11.9 to 11.10.

Video 43 This video shows white material being extracted from the foot and then entering the foot at another location. There is a brief appearance of black material prior to the white material entering the foot.

Video 44 This video shows a white particle exiting the foot and then re-entering the same foot. Consecutive frames before and after another the particle enters the foot are included.

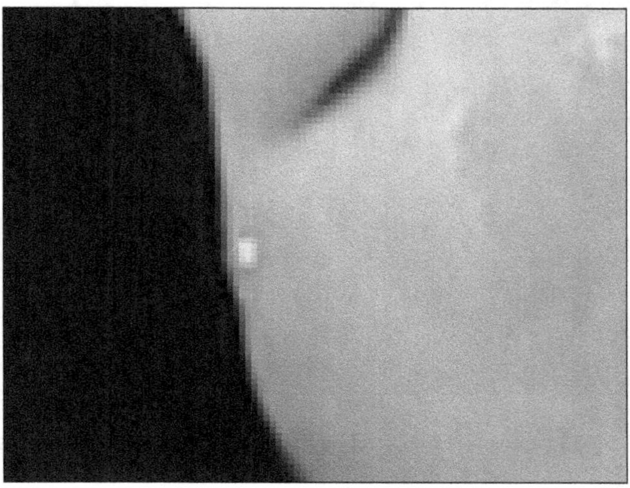

Fig. 10.5 This frame shows a particle that had just exited the same foot, immediately prior to the particle re-entering the same foot just distal to the exit point.

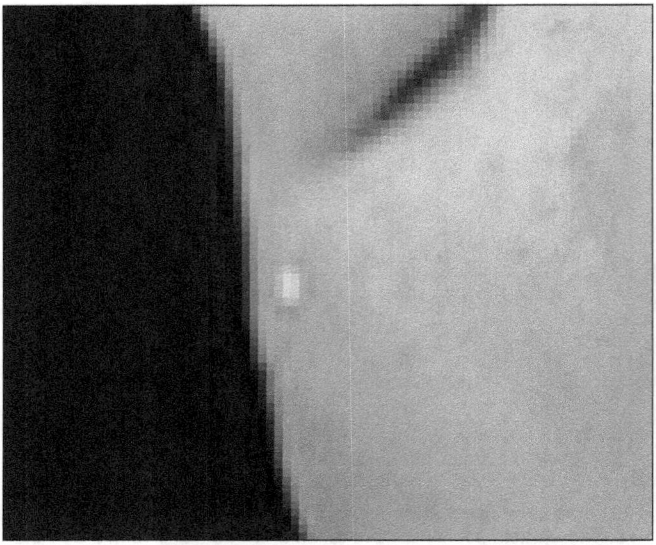

Fig. 10.6 This frame shows the same particle 33 ms later. The particle is 2 mm distal to its position in Figure 10.5.

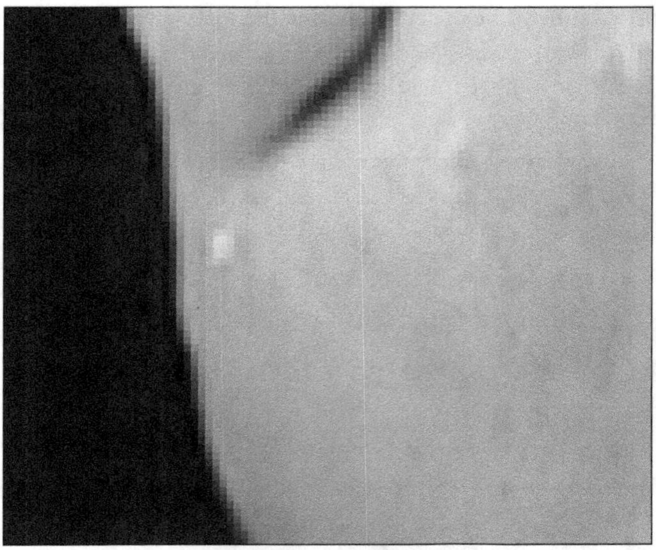

Fig. 10.7 After 33 ms, the particle has moved a further 2 mm distal from its position as seen in Figure 10.6.

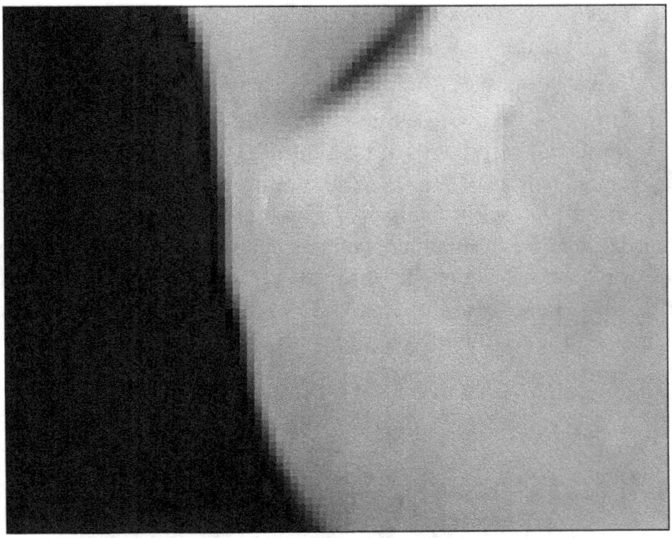

Fig. 10.8 After 33 ms more, the particle is significantly less visible than it was in Figure 10.7 while the position of the particle on the foot has not changed. The diminishing particle indicates that it is being absorbed into the dermis.

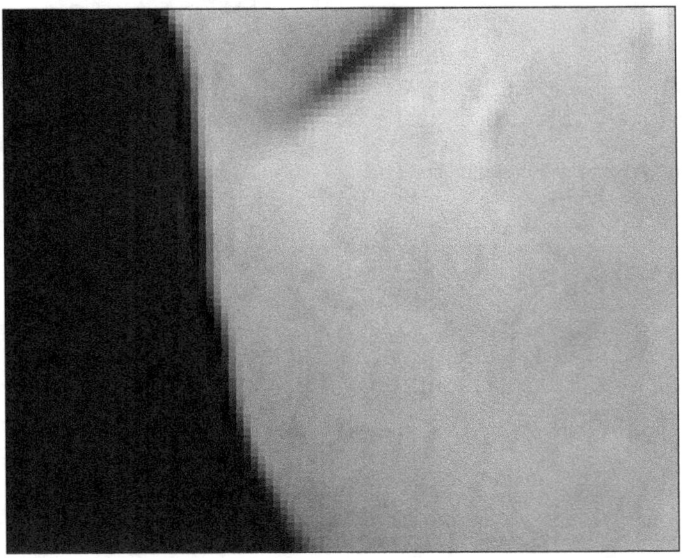

Fig. 10.9 This frame was taken 33 ms after Figure 10.8. Note that the particle has all but disappeared. A residual streak of white remains where no white streak was present before the particle landed on that spot.

Video 45 This video shows a white strand being extracted from the lower LD3 and then re-entering the same digit.

Video 46 This video shows a hair that disappears when disturbed by tweezers.

Video 47 This video shows a white piece of lint that disappears in 33 ms. The phenomenon was not recognized at first and was only noticed weeks later during a frame by frame analysis of the video. The initial interpretation was that lint was being brushed off the toe. In delusional parasitosis, in which diagnosis I include Morgellon's disease, the complainants often claim that lint is spontaneously exiting their epidermis. The subject of this study also complained of c-shaped stiff pieces of lint exiting his epidermis. This video showing that lint can enter the epidermis may explain why lint is contained in the epidermis of these individuals.

The following four frames are taken 33 ms apart.

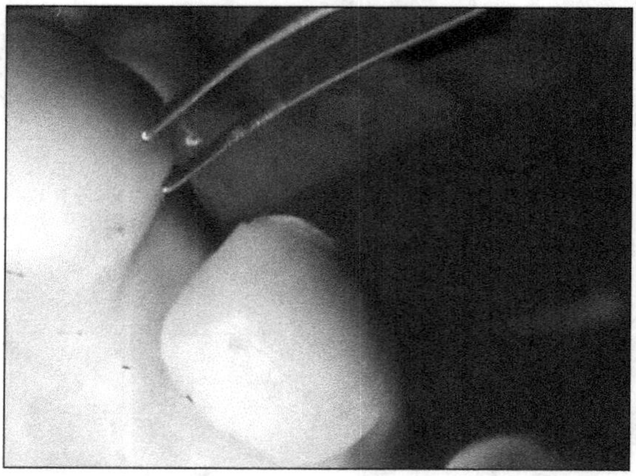

Fig. 10.10 A piece of lint is stimulated by a pair of forceps.

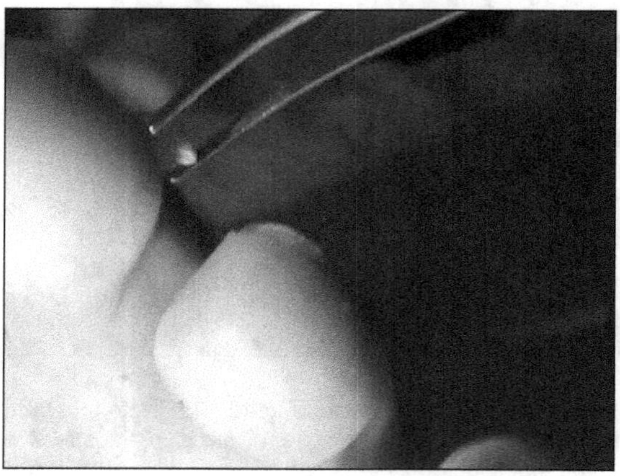

Fig. 10.11 The lint begins to move 33 ms later.

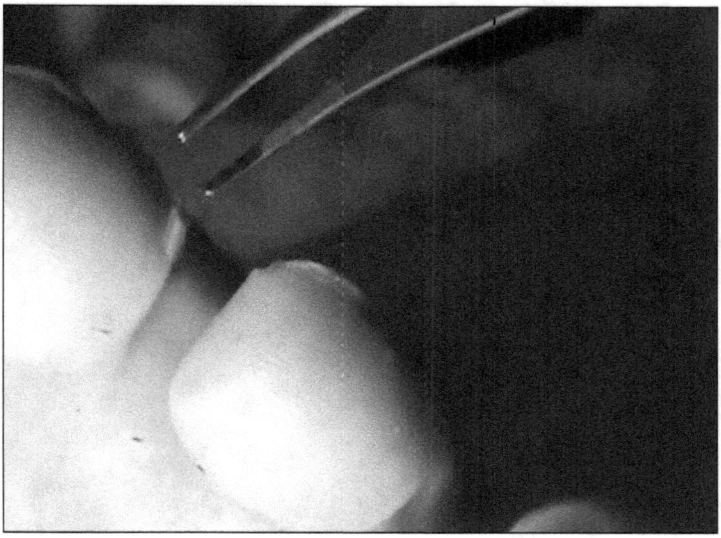

Fig. 10.12 The lint begins to descend at a rapid velocity.

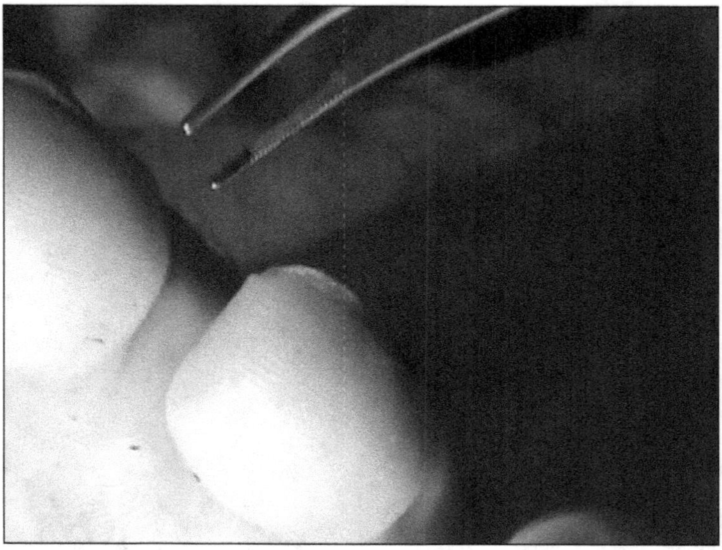

Fig. 10.13 The lint has completely disappeared within 33 ms. The lint should still be in the field of view unless it has been absorbed into the toe. Compare this to the frame by frame analysis in Figures 9.20 to 9.23 where a solid particle falling is visible for a few frames. Lint would have more drag and should fall even slower.

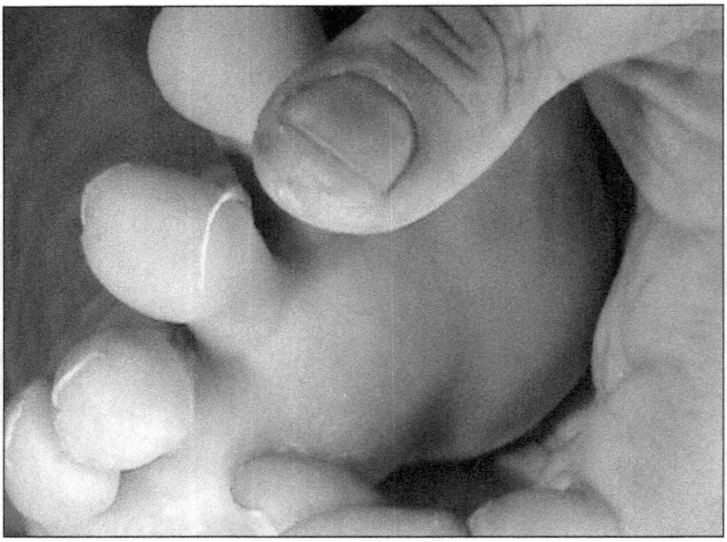

Fig. 10.14 A feather is found on the foot after the socks have been removed. On the suspicion that the feather may contain the organism, it was manipulated with the thumb and some of the results shown below.

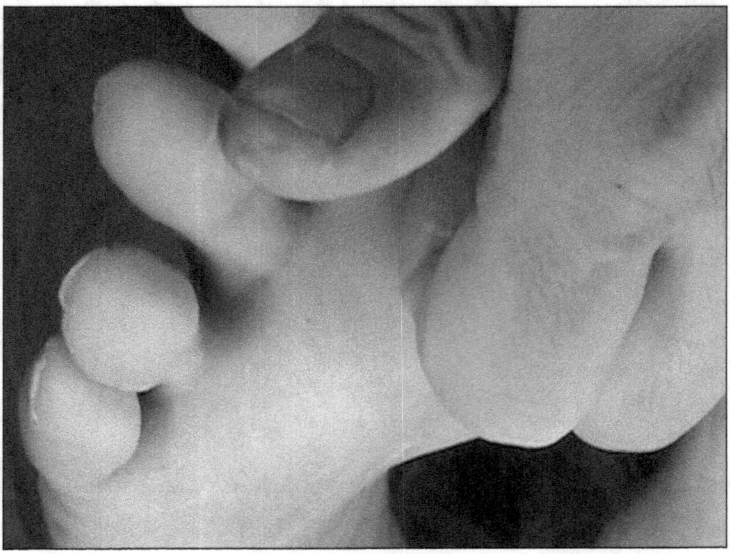

Fig. 10.15 The feather appears to be almost entirely imbedded within the toe.

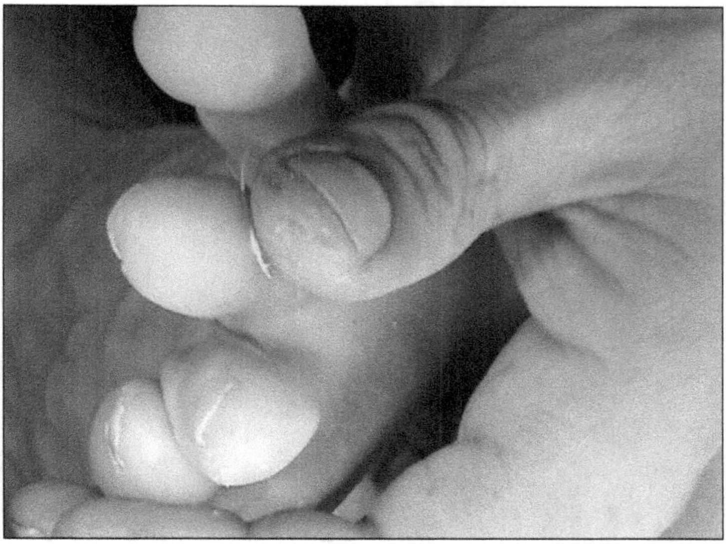

Fig. 10.16 The middle portion of the feather is embedded within the epidermis of the toe.

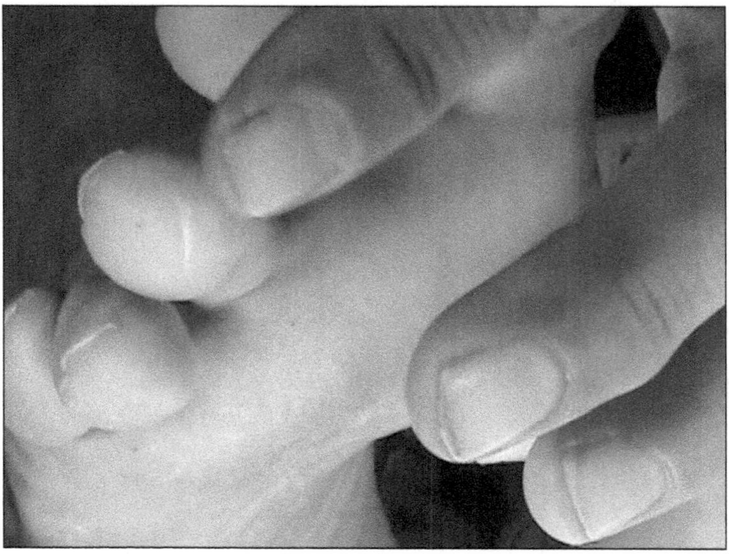

Fig. 10.17 Most of the dorsal portion of the feather is imbedded in the epidermis of the toe.

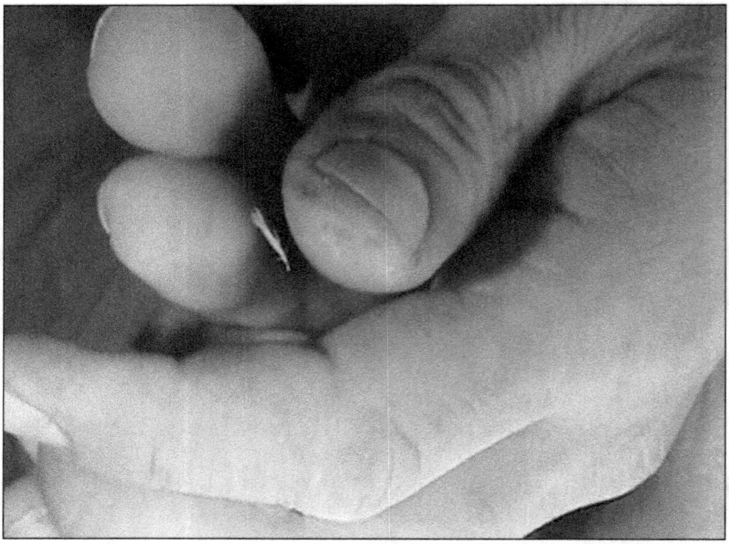

Fig. 10.18 The same description as Figure 10.17.

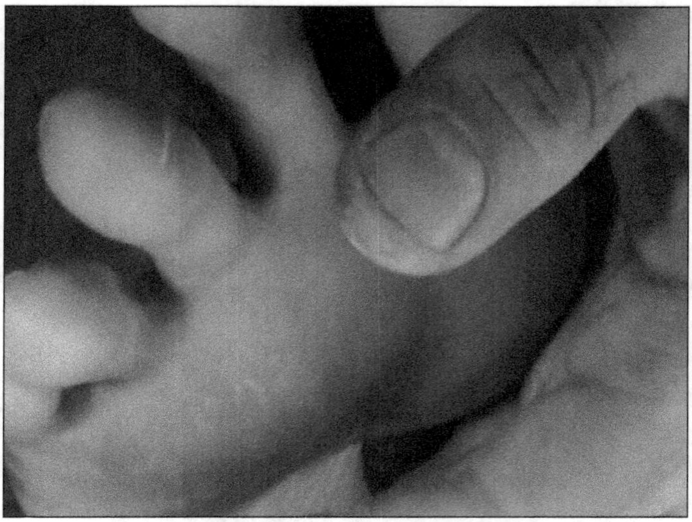

Fig. 10.19 The ventral portion of the feather is imbedded in the epidermis of the toe.

Video 48 This video shows a fluffy particle coming off the plantar surface.

Video 49 This video shows white particles exiting one foot and entering the other foot.

Video 50 This video shows a fluffy particle exiting the foot.

Video 51 This video shows various particles exiting the feet and flying in random directions.

11

The Microorganism Mobility – An Internal Propulsion System

The internal propulsion system is one of the unique properties of this microorganism. The propulsion system is most likely the mechanism utilized by the organism to seek new sources of cocaine as well as to travel through the body within the epidermis. How this propulsion system works is yet to be determined.

Video 52 This video shows two white particles going in different directions, reducing the possibility that wind or electrostatic attraction are responsible for the movement.

Video 53 This video shows a white particle exiting the foot and changing direction.

Video 54 This video shows white particles coming off the floor and flying.

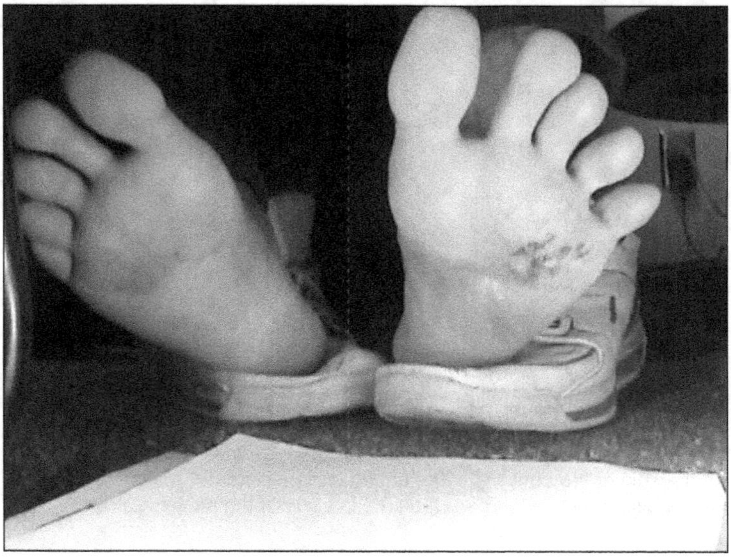

Fig. 11.1 Although difficult to visualize, this frame shows the particles as they start to move.

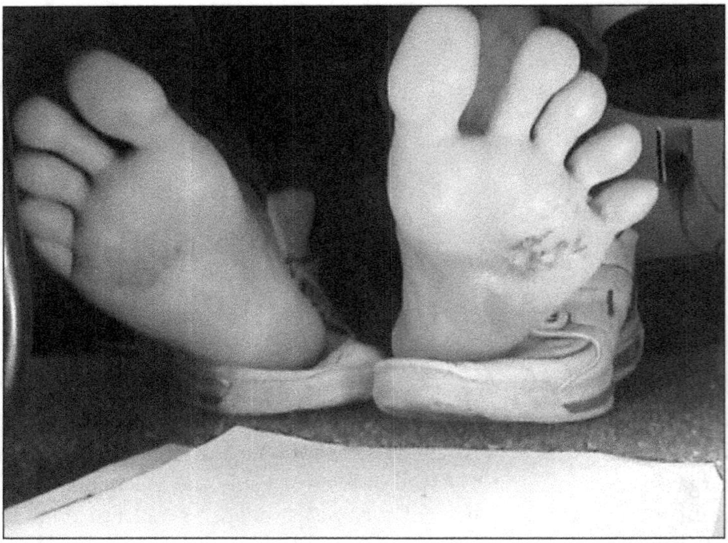

Fig. 11.2 This frame is taken 132 ms later. The particles are still difficult to visualize.

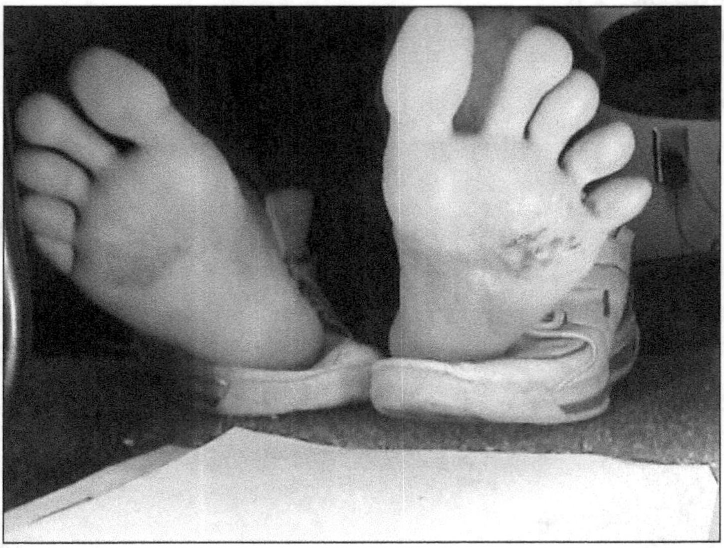

Fig. 11.3 This frame is taken 132 ms after Figure 11.2. One particle is seen just below the middle of the sole of the right shoe (the shoe on the left side of the photograph). The other particle is seen to the left of the paper (on the left side of the photograph).

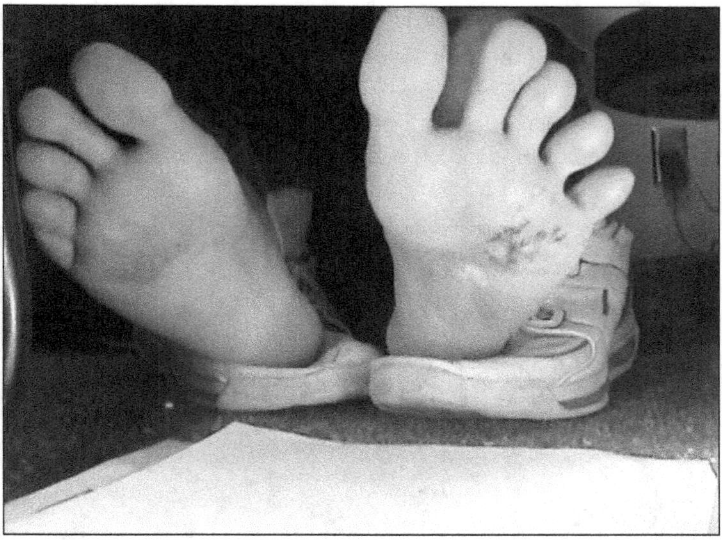

Fig. 11.4 This photograph is taken 165 ms later and shows one particle near the sole of the right shoe and only a small portion of the tail end of the other particle is seen on the far left of the photograph.

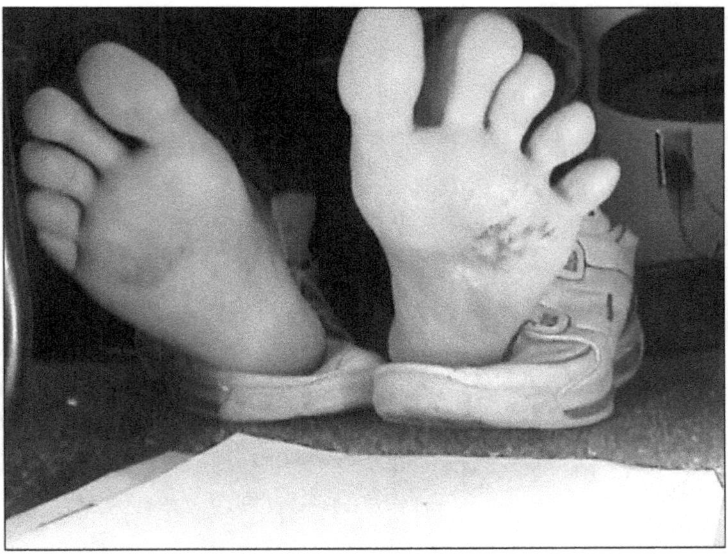

Fig. 11.5 After another 100 ms, one of the particles is seen heading towards the same area where the first particle went. A 300W halogen Welch-Allyn floor lamp is at the location where both the particles went.

Video 55 This video shows a particle flying from the bottom of the video.

Video 56 This video shows a particle entering and then exiting the lower lid.

Video 57 This video shows a white particle exiting the foot and then changing direction while similar particles fall directly towards the ground eliminating the possibility of wind currents initiating the change in direction. The following ten photographs are taken over a 1.5 second period.

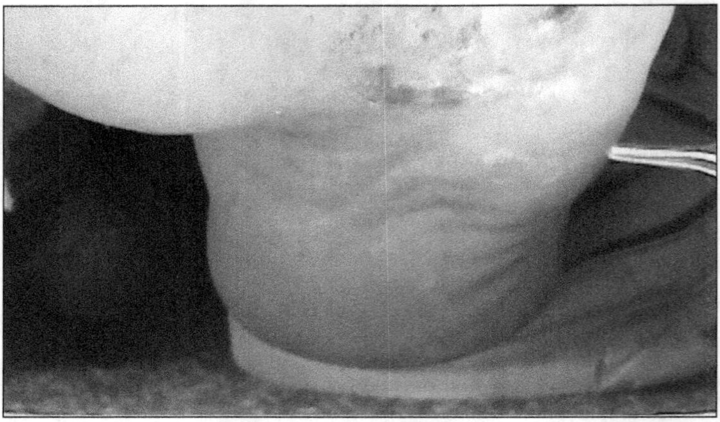

Fig. 11.6 The first particle is seen in the area of the distal border of the calcaneus. The particle is heading inferiorly and laterally.

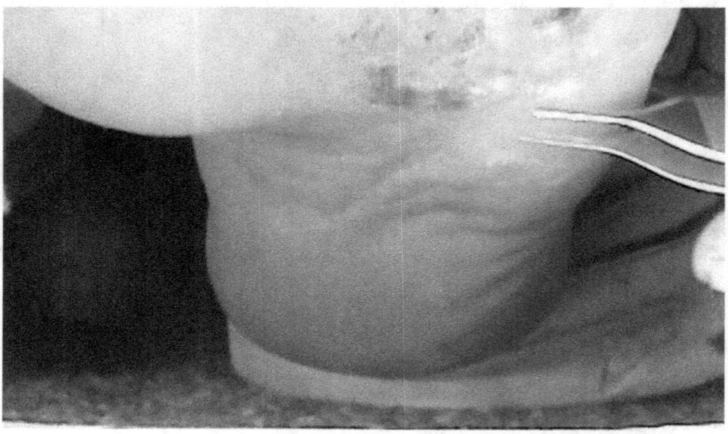

Fig. 11.7 The particle is seen a little lateral and inferior to the last photograph 200 ms later.

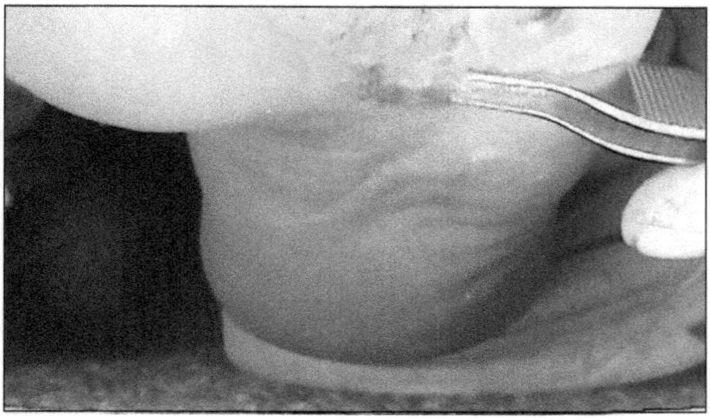

Fig. 11.8 The particle is now moving laterally (seen 200 ms after Figure 11.7).

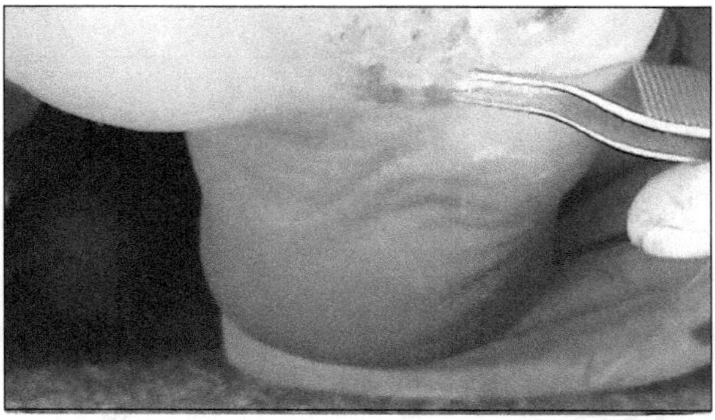

Fig. 11.9 There are now two particles 66ms later. One continues to move laterally while the other is moving inferiorly.

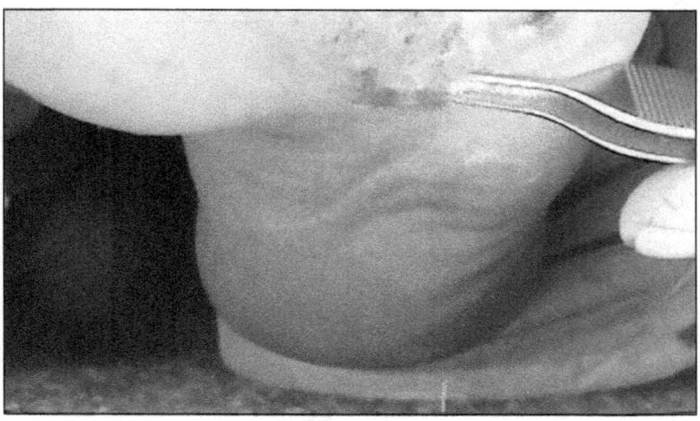

Fig. 11.10 The particles continue in their same respective directions as seen 200 ms after Figure 11.9.

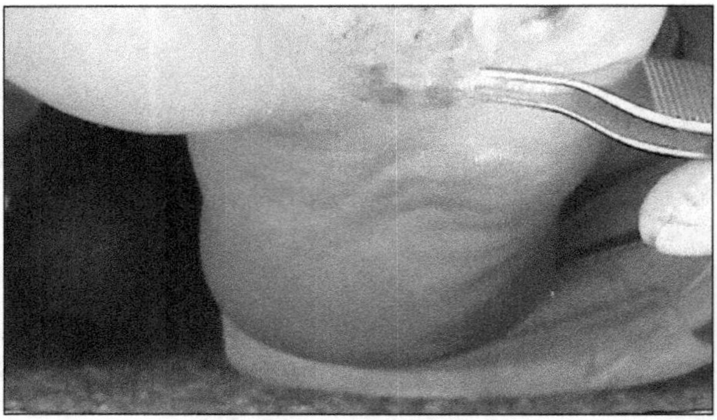

Fig. 11.11 The particle is now traveling medially 333 ms later. This represents two direction changes: the first direction change was from an inferior-lateral direction to lateral direction and the second direction change was from the lateral direction to a medial direction.

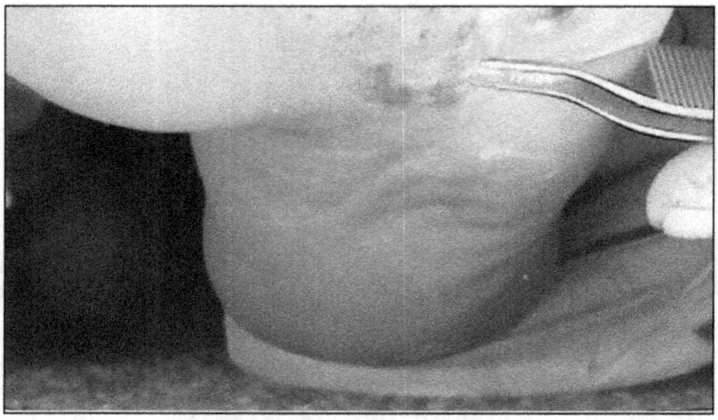

Fig. 11.12 The particle continues to move medially.

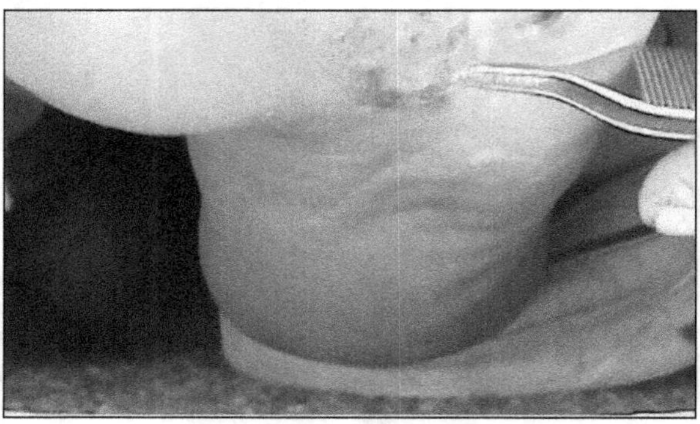

Fig. 11.13 The particle continues to move medially.

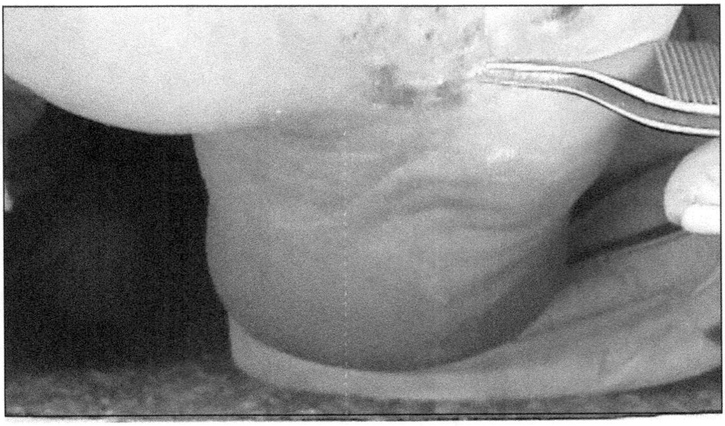

Fig. 11.14 The particle continues to move medially.

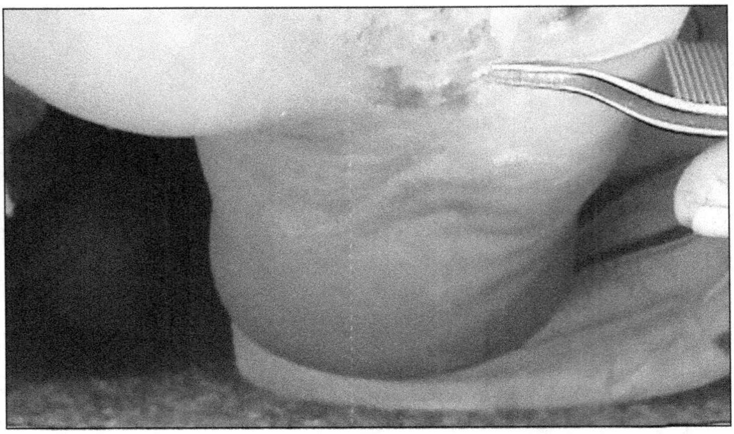

Fig. 11.15 The particle is now moving slightly superiorly and medially. Note that the particle is now a strand.

Video 58 This video shows white particles flying around. The paper on the floor does not stir.

Video 59 This video shows white particles flying towards the light source. The paper on the floor does not stir indicating that the particles are not propelled by wind currents.

12

Green Feet Illuminate the Role of the Epidermis in the Life-Cycle of the Microorganism

While extracting white material from the feet, occasionally a small, flat, hard piece of keratinized epithelium was removed from beneath the epidermis. On the hunch that perhaps the sub-epidermal plates of keratinized epithelium were recruited from the plantar surface of the feet, both feet were colored green with a permanent magic marker. The subject was subsequently surveyed for the color green, the theory being that the green epithelium would be recruited by the microorganism and its movements tracked by a trail of green color. Initially, the hypothesis was that the ectopic green color would be found close to the feet. It was surprising, then, to discover green material emerging from the eyes (see Figures 7.6, 8.2, 8.3, 12.1 and 12.2).

Video 20 Although this video was already presented in Chapter 5, it is revisited here to demonstrate the movement of the green color that was applied to the feet (see Figure 4.7).

Video 60 This video shows dozens of white particles covering the eye globe. Careful observation reveals that some of these particles appear and disappear.

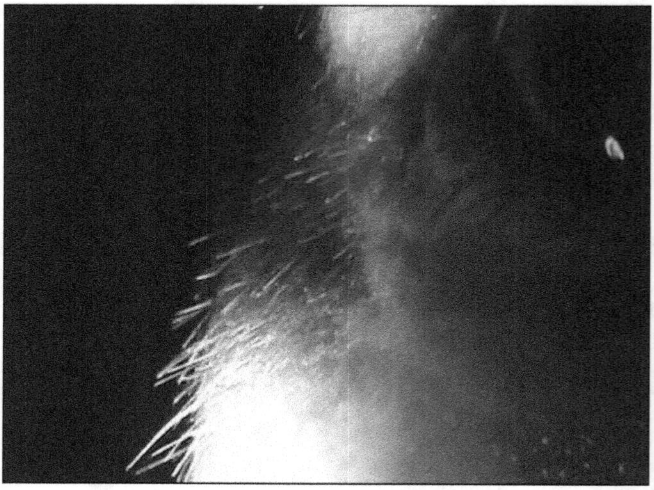

Fig. 12.1 This frame shows that many of the hairs inferior and lateral to the right eye may be luminescent while being illuminated by a Welch-Allyn 46000A headlight. Some of the hairs shown here have luminescent white particles attached. A green-tinged plaque of translucent material is seen inferior to the lower right eyelid (see DVD for color photograph). Note that the feet were previously covered in green ink to monitor the role and movement if any, of the epidermis in the life cycle of the organism.

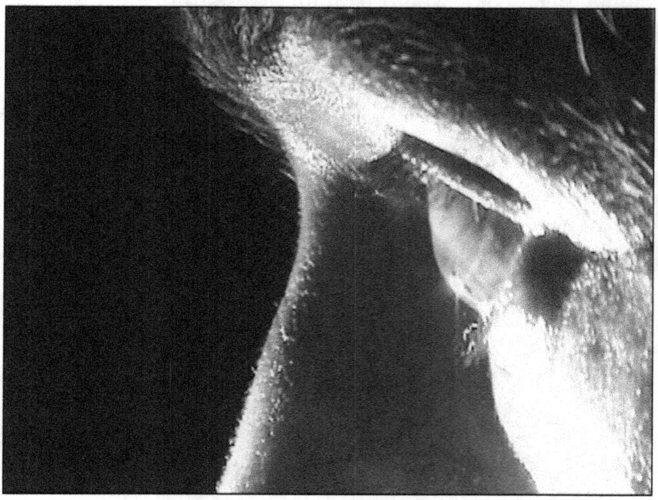

Fig. 12.2 The superior sclera has a green tinge under illumination with the Welch-Allyn headlight. The plantar surfaces of the feet were previously covered in green ink to determine the role of the epidermis in regards to the micro organism. A blue-tinged band is seen extending from the lateral upper eyelid margin to the pupil and several of the eyelashes are coated in a reflective and or luminescing white material (see DVD for color photograph). The blood vessels normally seen covering the sclera are not visible suggesting the presence of a foreign substance covering the sclera.

13

Coated Hairs under the Microscope

A paroxysm of pruritis on the dorsal aspect of the forearm corresponded with the appearance of minute white particles on the hair follicles in the same area (see Figure 13.1). Some of these hairs were scraped onto a slide with a scalpel and examined under a microscope. At 400X magnification, the white particles had a flat crystalline appearance (Figures 13.2, 13.2, 13.5 and 13.10) and under the Welch-Allyn 46000 REVA light they appeared to have areas of luminescence (see Figures 13.4, 13.6, 13.9 and 13.11).

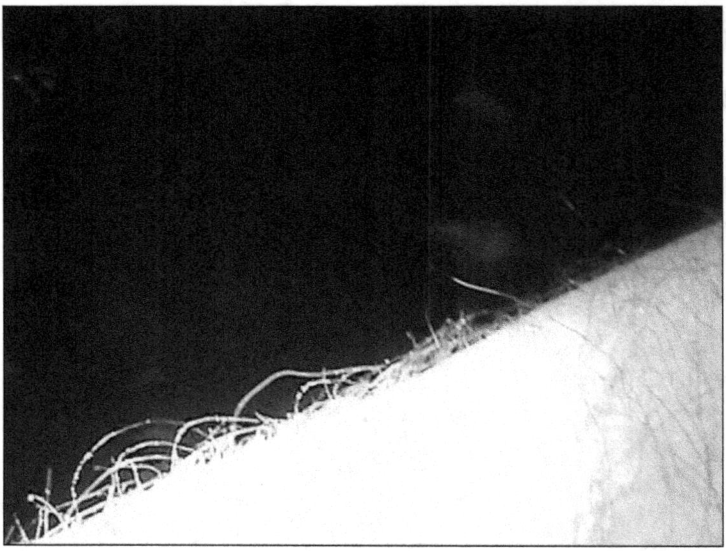

Fig. 13.1 A pruritic patch of dermis on the dorsal aspect of the distal upper limb illuminated with a Welch Allyn 46000A spotlight. Note the hair follicles are coated with what appears to be a luminescent white material emitting green, blue and purple light (see DVD for color photograph).

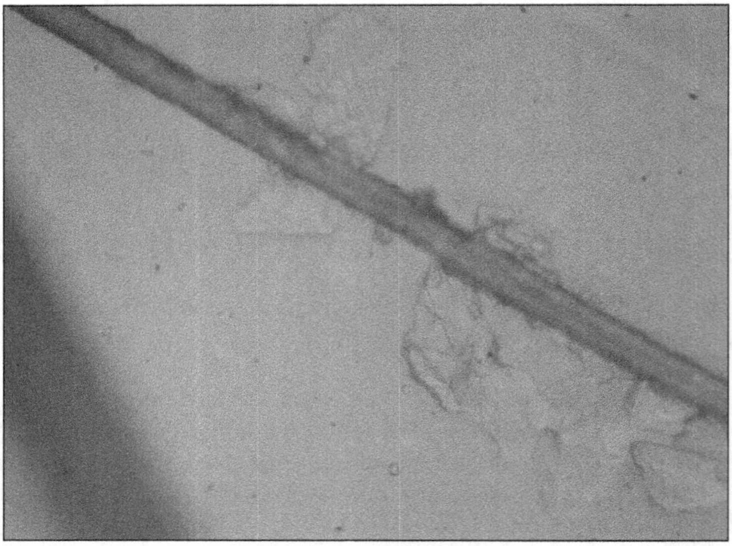

Fig. 13.2 This photograph shows one of the hairs from the forearm as seen in Figure 13.1, at 400X magnification. The white material seen on the hair follicles without magnification most likely correspond to the crystalline materials seen with the microscope. This material will most likely be determined to be keratin recruited by the organism from the epidermis.

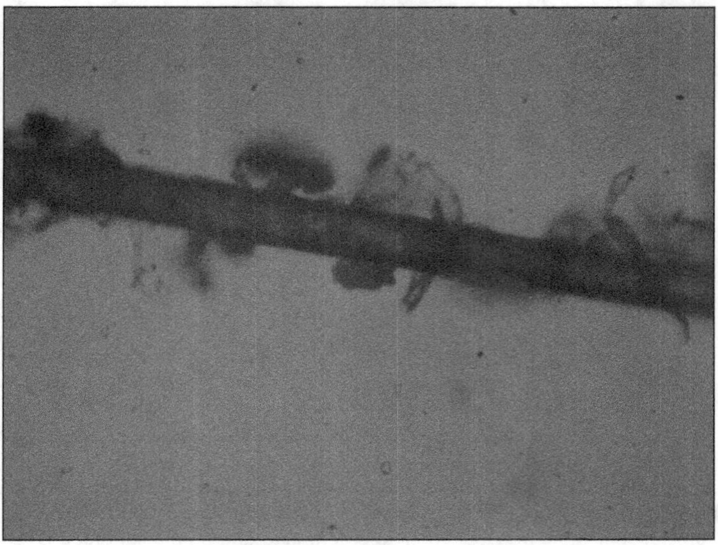

Fig. 13.3 A hair follicle scraped from the dorsal aspect of the distal upper limb under 400X magnification with normal lighting.

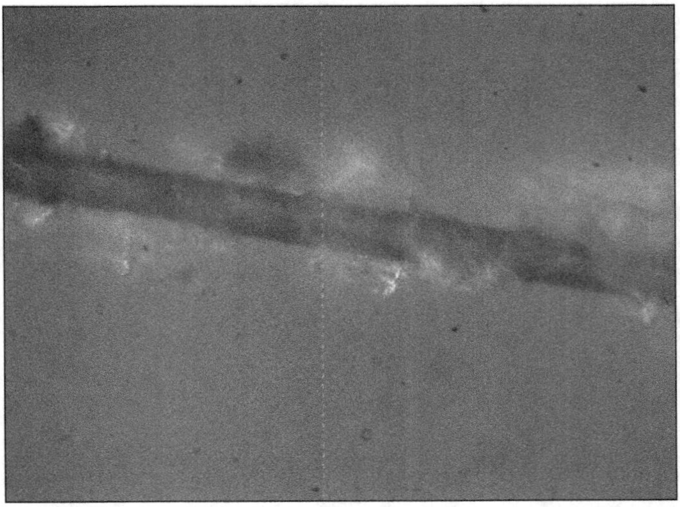

Fig. 13.4 The same hair follicle illuminated using the Welch-Allyn 46000 REVA headlamp. Note the areas of luminescence correspond with the areas of filamentous materials as seen in Figure 13.3 when using only the normal incandescent aperture lighting provided with the microscope. Also worthy of noting is that the luminescent points of light located on the dark filaments seem to illuminate the dark filaments demonstrating that they consist of a complex network of filaments not unlike the structure of a bush with braches and twigs.

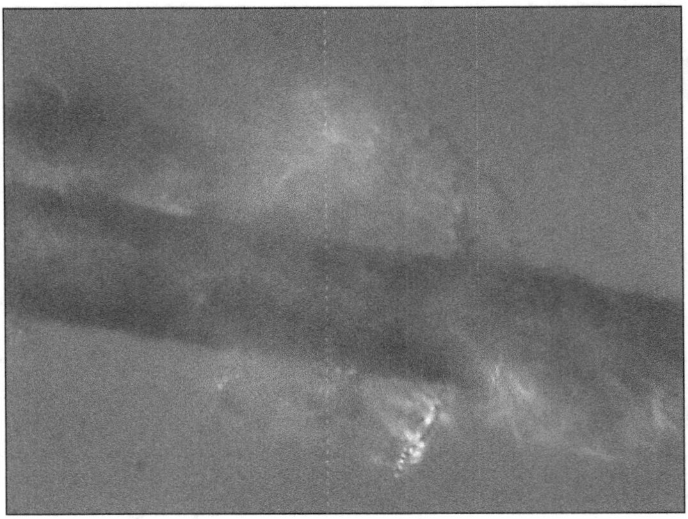

Fig. 13.5 This is a closer view of the hair seen in Figure 13.4 showing the filaments and lights in more detail. The multicolored lights are located along dark filaments that are similar to microscopic trees growing out of the hair follicle in appearance. The filaments have the same appearance as the filaments seen in the specimen extracted from the foot (Figures 9.26 to 9.29).

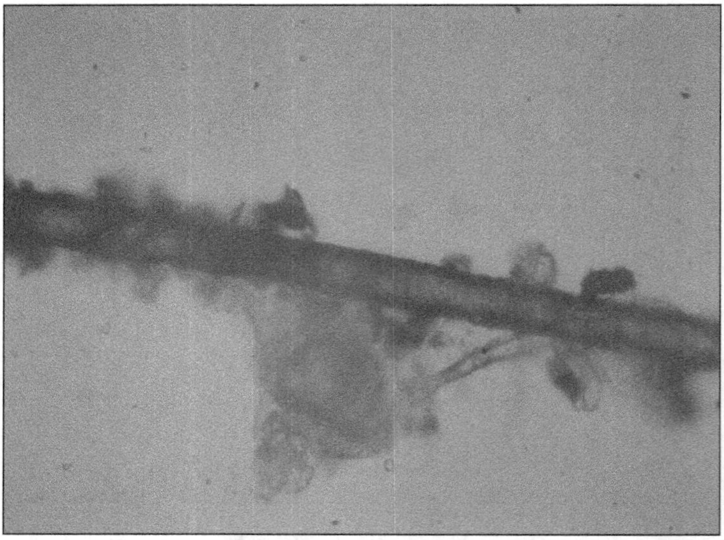

Fig. 13.6 This photograph shows another hair follicle as seen at 400X magnification.

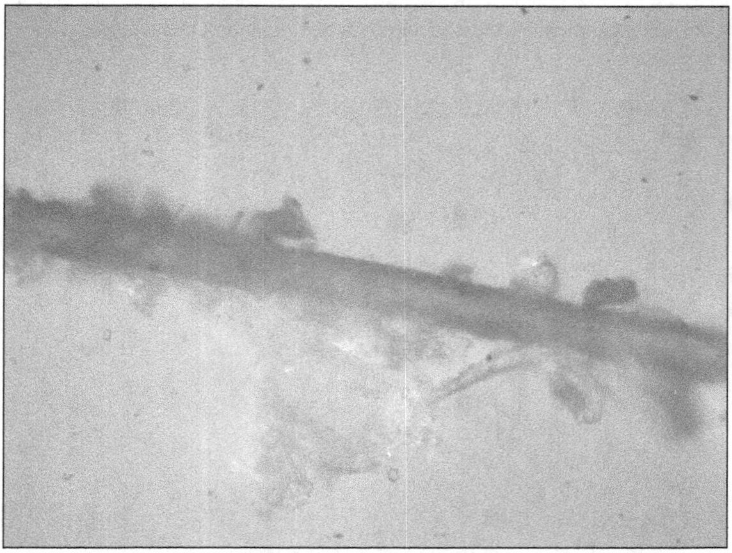

Fig. 13.7 The same hair follicle illuminated with a Welch-Allyn 46000A headlight.

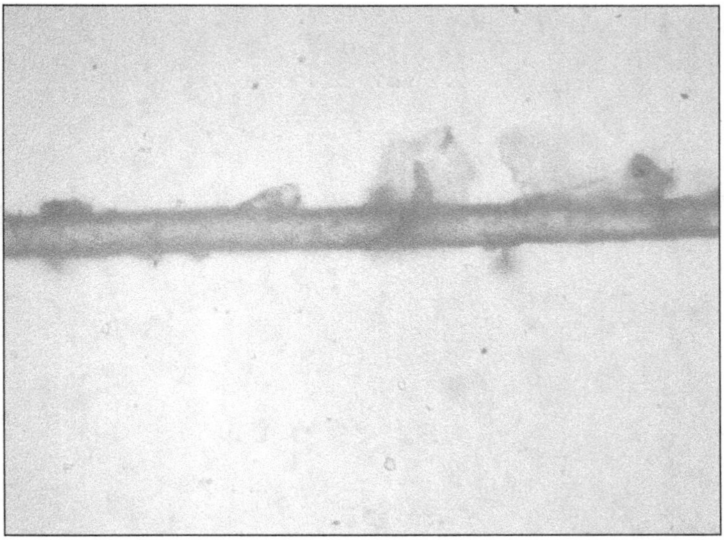

Fig. 13.8 This is a photograph of another one of the hairs scraped from the forearm.

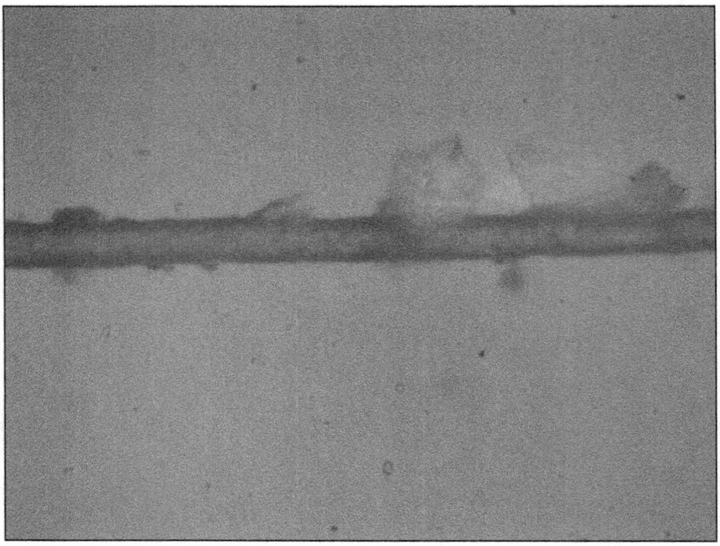

Fig. 13.9 The same hair as in Figure 13.9 but illuminated with the Welch Allyn headlight.

Fig. 13.10 The same hair again illuminated with the Welch-Allyn headlight. Note the area of strong luminescence just to the left of midline on the upper surface of the hair follicle.

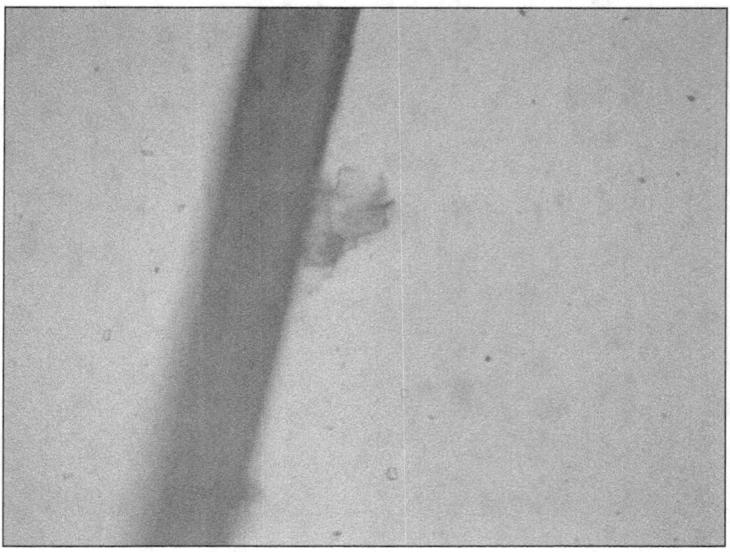

Fig. 13.11 Another hair from the forearm under normal aperture lighting showing another crystalline structure attached to the shaft.

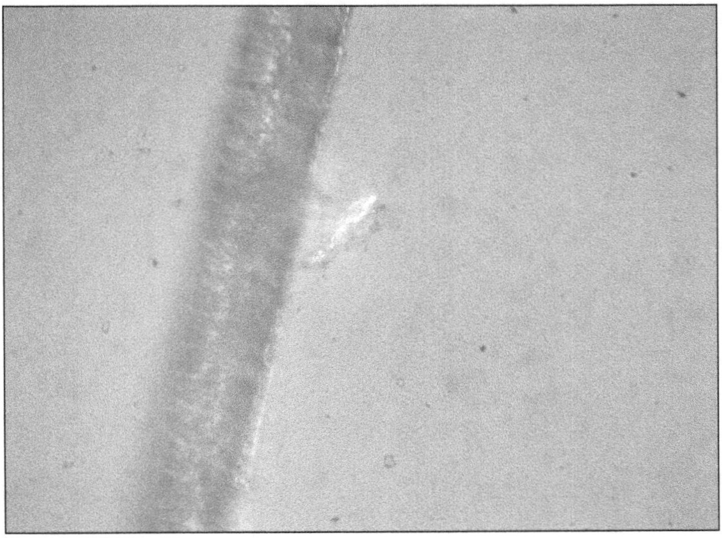

Fig. 13.12 This photograph shows the same hair follicle as seen in Figure 13.11 except that it illuminated by the Welch-Allyn headlight. Again, note the areas of pinpoint luminescence. Another observation of unknown significance is that the pinpoint luminescence extends along the entire strand of hair on the right side. This may indicate that the microorganism extends along the whole length of infected filaments probably explaining how the whole strand of hair enters the epidermis instantly.

14

The Micro Organism Possibly Growing on a Piece of Cocaine

The theory that crumbly cocaine bicarbonate may be infected with a micro organism prompted the examination of such a piece of cocaine, also known as crack cocaine. The results are shown below. Although far from conclusive, this may represent a "coca mite." Compare the appearance of this material to the material that came out of the foot as seen in Figure 9.29.

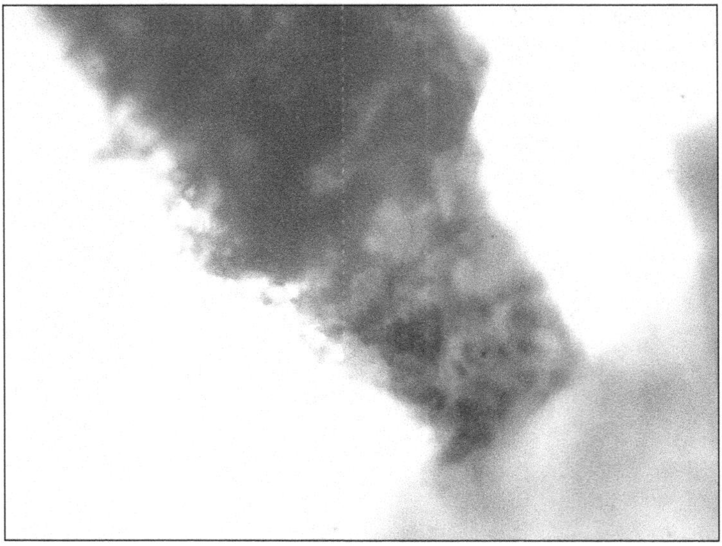

Fig. 14.1 A small piece of cocaine bicarbonate (bottom right) magnified at 100X shows an unidentified object attached to the surface.

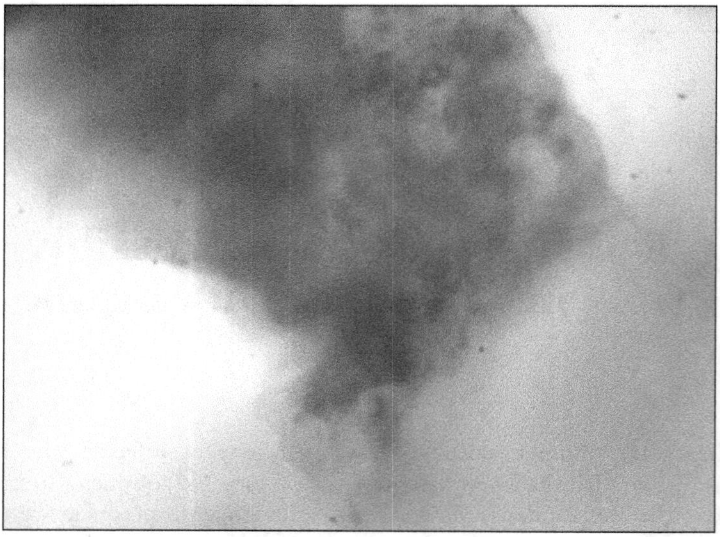

Fig. 14.2 The same object as seen at 200X magnification. The object may be attached to the cocaine bicarbonate along the entire interface however it may also be anchored to the cocaine at the inferior middle of the photograph where some of the object appears to be infiltrating the piece of cocaine.

15

A Summary of the Observations

The observations detailed above are best explained by the existence of a new form of microscopic organism, however they are equally explainable by a known microscopic organism with several, until now, undocumented properties. For the purposes of keeping this discussion simpler, the organism, new or existing will be referred to as a new microorganism. The interpretation of this investigation will begin with a summarization and analysis of the phenomena seen in the videos and photographs. The microorganism, new or existing, has several distinct properties which are explained below.

The microorganism is able to easily enter and exit the epidermis, cornea and sclera. There are numerous examples of this phenomenon described in the observations. Video numbers 19, 20, and 44 are just a few examples. The microscopic examination of affected hairs show a crystalline substance – most likely to be keratin recruited from the epidermis of affected individuals - surrounded and infiltrated by a filamentous material which is presumably the new micro organism. These sharp-looking crystalline materials may partially explain how the affected hairs enter the epidermis and eyes so quickly.

The micro organism can rapidly repair the damage caused by its entry and exit through the epidermis and optic globe. Video number 20 shows that the optic globe can be lacerated from top to bottom and yet no vitreous humor exudes. In an effort to extract the organism from the foot, holes were bored through the dermis with needle forceps. It became immediately apparent that the organism was sealing off the holes bored into the dermis by repairing the damage with a white material, presumably the keratin - micro organism complex. This phenomenon alone has enormous pharmacological potential perhaps in wound or scar management.

The micro organism may have antibacterial properties. Throughout a 15 month span of boring holes through the subject's feet with a dirty pair of needle forceps and white material and dirty hair entering the globe and dermis, the subject did not contract any other infections in the feet or eyes.

The micro organism attaches itself to mammalian hair, possibly avian feathers, cotton lint and other, as of yet unknown, materials such as cigarette tobacco, and other textile constituents. Perhaps this has evolved to provide the microorganism with a vector of infection and perhaps as an area for launching its spores – once the cocaine vapors are detected by the organism's chemo receptors - into the surrounding space to find other hosts or a cocaine supply. Think of an infected hair as the micro organism's version of the space shuttle and launch pad, while the vaporized cocaine is the launch ignition switch. In that analogy, the micro organism would represent the rocket fuel, engines and astronauts.

The attachment of the organism to other materials could explain the unrecorded phenomenon of the white c-shaped objects exiting the dermis. Figures 10.10 to 10.13 suggest that the microorganism may attach itself to strands of cotton that allows it to enter, and by corollary, exit the dermis. The piece of cotton was videotaped entering the epidermis. At the time that the subject reported the appearance of the c-shaped objects exiting his body through the dermis, he was sleeping in the basement with a white cotton blanket. It is possible that the microorganism was attached to the cotton fibers of the blanket and these fibers were then entering the epidermis of the subject for days or weeks. Then, for some unknown reason, the fibers left or were expelled from the body within a short period of time.

Another possible explanation for the entry of white cotton fibers into the epidermis is that the microorganism is in the epidermis and it absorbs cotton fibers into the epidermis. In this scenario, the cotton fibers do not have to be infected with the organism nor do they necessarily have to be infiltrated by cocaine molecules. There would have to be some reason the organism is attracted to any kind of fiber, perhaps these fibers are used by the microorganism as vectors.

There are several possible explanations that can account for the expulsion of the c-shaped pieces of lint from the subject's body. The first explanation is that the soak in aqueous Sunlight laundry detergent somehow initiated the subject's body to expel the lint from the dermis. Another possible explanation is that the microorganism was stimulated to leave the body by a cocaine stimulus outside the body. Another possibility is that most of the microorganism was killed by exposure to the sunlight detergent, thereby enabling the body to expel the lint. Still another explanation, and this one is the most favored by the author, is that the Sunlight detergent somehow neutralized the cocaine in the epidermis and the microorganism-lint complex expelled itself from the subject's epidermis in search of more cocaine, possibly attracted by a small concentration of cocaine in the environment outside the body.

The above theory would also explain one of the symptoms reported by other users of vaporized cocaine. The symptom of scalp tingling followed by the expulsion of fibrous materials from their scalps reported by cocaine users could be caused by the same mechanisms that explained how the c-shaped pieces of lint were exiting the epidermis of the subject. Basically, the infected fibers are stimulated by cocaine vapors in the surrounding environment to exit the epidermis.

In Chapter 1, the University of Toronto professor of infectious diseases and one of the world's leading authorities on the so-called "coca mite" stated that the specimens brought to him by cocaine users turned out to consist of lint. According to complainant reports, the lint spontaneously exits the scalp shortly after ingesting vaporized cocaine. The theory of why the user of cocaine has lint falling off their scalps is outlined below.

The first step to explaining this phenomenon is to explain how lint gets into the scalp. There are a couple of theories that could explain this phenomenon. The first theory, and the one favored by the author, is that the cocaine user's clothing and bedding becomes infected with the microorganism by being exposed to the microorganism from cocaine crumbs or from particles exiting from the user's epidermis. Figures 10.10 to 10.13 shows that cotton enters the epidermis in the same manner as the white particles and the hair follicles enter. Therefore, the microorganism infects the cotton fibers, which then allows those cotton fibers to enter the epidermis. The second theory is that the microorganism living in the epidermis absorbs any kind of fibrous material that contacts the epidermis in order to utilize it for some purpose, perhaps to use the fibers as a vector of transportation.

The cotton fibers then work their way to the scalp using their endogenous propulsion system possibly in a keratin-microorganism matrix but equally they could travel on their own.

Once in the scalp, the cotton fibers congregate into clusters which are then stimulated by vaporized cocaine fumes emitted from the cocaine smoker's pipe into the surrounding environment including the scalp. The cocaine stimulated cotton fibers then propel themselves out of the epidermis over the scalp and these clumps of lint are then reported as being parasites that exited the scalp. An examination of these clumps of fibers would reveal that they consist of pieces of lint.

The microorganism seems to be attracted to halogen lights. The very first abnormality, remembered retrospectively, was the white particles flying into the halogen pot lights located in the basement of the research subject's residence. Although primarily thought to be baby spiders, in retrospect, these particles are now thought to be the microorganism. Some of the particles seen traveling into the pot lights resembled minute textile fibers, supporting the theory that the microorganism infects textile fibers as well as animals. Videos of the feet demonstrate that the white particles that exit the epidermis of the feet often fly towards the 300 watt Welch-Allyn halogen floor lamp as seen in Chapters 9 and 11.

The microorganism changes its shape while traveling through the air. As seen in Figures 5.17 to 5.24 and 9.1 to 9.9, the particle elongates and contracts while traveling through the air. In Figures 5.17 to 5.24 the particle elongates 2.5 times its former length prior to disappearing into the epidermis of the premaxilla. The microscopic structure of the organism as seen in Figure 9.28 is consistent with this phenomenon. The microorganism consists of a filamentous matrix and the larger filaments within this matrix are universally furled. When the particle is seen to be elongating, the filaments are most likely simply unfurling in response to some sort of as yet undetermined stimulus, possibly cocaine or possibly a stimulus emitted from the host's epidermis.

Once in the host, the microorganism probably recruits epidermal cells and/or keratin from the epidermis and then incorporates it into its filamentous matrix. The most likely explanation for this is that the microorganism uses keratin as a protective barrier between it and the host's immune system. The keratin is apparently also used in the repair of the damage done to the host's dermal system by it and any other mechanism such as debridement with forceps. The evidence for the presence of keratin in the microorganism's external matrix is simple and is described in the following section.

Some of the material extracted from the feet was examined under a microscope and photographs were taken. These were then examined under a microscope and determined to be keratinized epithelium (see Figures 9.12 and 9.13). However, there appeared to be an abnormal amphorous substance containing circular particles and a black filamentous material present in the epithelium. At this point it was necessary to determine the origin and role of the amphorous substance in the epithelium of the feet in the phenomena associated with this organism. In order to do this a simple experiment was devised. The plantar surfaces of both feet were stained with a green colored permanent marker with the hope that the movement of the epithelium could be tracked and monitored (see Figure 4.7).

Further videos were monitored for the presence of green colored material. Several videos showed that the green materials were then present in excretions from the eyes (see Figures 7.6, 8.2 and 8.3). This supports the hypothesis that the micro organism is absorbing keratin from the feet creating a keratin-micro organism-matrix and this matrix then moves through the body within the epidermis eventually traveling through the eyes. Once in the eyes, some of the keratin-micro organism matrix is secreted and the green color becomes apparent in the periorbital region.

With itself surrounded by keratin, the keratin-organism complex moves through the body like a worm, however, I am almost certain that it does not go deeper than the dermis. I believe this is so because the progression of the micro organism can usually be slowed or halted completely by compression of the toes resulting in grossly swollen feet (see Figures 9.10 and 9.11). Release of the toe compression alleviates the swelling in less than 30 minutes.

The micro organism has a propulsion system that allows it to exit and enter the epidermis at relatively high velocities as well as to accelerate and change directions outside the epidermis. This propulsion system also propels it and the amphorous keratin-micro organism matrix through the epidermis as it travels through the body.

The micro organism probably consumes cocaine which is one of the most deleterious substances on earth (obviously way behind carbon dioxide). The videos of the spontaneous expulsion of materials from the feet were all made immediately after vaporized fumes from cocaine were blown onto the feet. Moreover, other members of the household of the cocaine using subject were not symptomatic and therefore, it appears that only persons with cocaine in their systems are infected with this organism.

The cocaine is probably used by the organism to avoid detection as this substance is a well-known strong analgesic and vasoconstrictor. Areas of the feet where the amphorous substance was present were pain-free even when boring a hole completely through the dermis and into the subcutaneous tissues with needle forceps. However, if the needle forceps contacted an area of the plantar surface where the micro organism was not active, the subject would experience a tremendous amount of intense pain.

The utilization of cocaine by the micro organism is probably its most important characteristic. Cocaine has destroyed millions of lives, diminished whole nations and costs the world economy trillions of dollars annually. If this destructive substance could be eliminated from the earth, perhaps through the genetic engineering of this microscopic organism, we could be free from a serious epidemic that is affecting the entire world.

The question arises here is why do non-cocaine using persons are also labeled with the diagnosis of delusional parasitosis. The answer may be either that cocaine makes the organism more virulent or that non-cocaine users are infected with a different strain of microorganism.

The micro organism is luminescent under certain light conditions namely, using a Welch-Allyn 46000 REVA headlight. The keratin crystalline materials seem to refract and reflect the light, however, hundreds of tiny points of luminescence are located along the black filamentous materials surrounding and infiltrating the white material. As some fungi are known to be luminescent this organism is possibly some sort of fungus.

16

The Structure of the Microorganism

Determining the structure of this new micro organism presented several difficulties. First and foremost was that the micro organism appears to utilize materials absconded from the infected individual as well as possibly utilizing materials recruited from its surrounding environment. It goes without saying that the process of separating what constitutes the organism from what is extra-microbial is impossible without knowing more about it. Having that said, what follows is an attempt at discerning the structure of the microorganism while keeping this inherent difficulty in mind.

Microscopically, the organism appears to have up to three forms. The first possible form is a dark filamentous strand or fiber seemingly attached to and surrounding affected hair follicles and also found within the white material contained in the feet below the epidermis. However, it is not yet known whether these filaments are part of the organism or whether they are materials recruited by the organism and integrated into the microscopic organic structure.

Associated with the hair follicles are crystalline structures that are not luminescent under the aforementioned lighting conditions. The crystalline appearance could explain how the organism enters and exits the epidermis painlessly and rapidly. Figures 13.2 and 13.3 show that the white materials found on the coated hairs are sharp-looking. These crystalline materials found on the hair shafts are probably related to the white material contained in the keratin-micro organism matrix seen coming out of the feet and seen in Figures 9.26 to 9.29.

The third possible form of the organism is the luminescent material that is seen under microscopic examination. Figures 13.4, 13.6, 13.9 and 13.11 all show that there are hundreds of miniscule points of light, indicative of the presence of luminescent materials, seen on the filaments. These could be the organism's spores or the only form of the organism. In this scenario, the white materials, both crystalline and amphorous, and the dark materials would be recruited from the host and somehow utilized by the micro organism.

Another possibility is that there is more than one organism working in some form of symbiotic relationship. Although the existence of more than one organism is highly unlikely, this possibility is mentioned only because anything is possible.

It is my conjecture that the dark filamentous material constitutes the body of the micro organism while the crystalline material is made up of keratin recruited from the epidermis by the micro organism possibly as a source of nutrition, mobility and/or invisibility. The luminescent points are probably also part of the organism. Whether these miniscule entities are going to eventually be determined to be an integral and inseparable part of the main functioning organism, or whether they are solely the spores of the micro organism, is difficult to predict at this time.

The material exiting the epidermis of the feet appears to be macroscopic versions of the material associated with the hair follicles. Figure 13.3 shows that the organism forms a complex matrix of a dark filamentous superstructure surrounding and surrounded by crystalline white material that is responsible for the shape and luminescent properties of the particles. The reflective white materials of the hair follicles probably correspond to the reflective areas of the macroscopic materials exiting the feet. The filaments attached to the hair follicles correspond to the dark filamentous matrix seen in the macroscopic white materials exiting the feet.

17

A Model Accounting for the Observed Phenomena

The model that best accounts for these observations is a mycotic or fungal microorganism. The microscopic appearance at 400X magnification shows a possible filamentous growth originating on several different hair follicle shafts. There is also a filamentous matrix located in the white materials that exited the feet. Therefore this micro organism is probably not an animal. The microorganism lives within the dermis and probably does not rely on photosynthesis and therefore is probably not a plant.

There are hundreds of points of luminescence seen along the dark filaments in both on the hair and in the micro organic matrix extracted from the feet. Luminescence under certain lighting conditions is a characteristic shared by mycotic organisms.

At this point, it is acknowledged that the interpretation of the observations presented here may be controversial to some. On the whole, the totality of the evidence presented here and the videos available on-line, in this opinion, clearly demonstrate several as of yet undocumented phenomena. Although specific video sequences may possibly be explained by previously known biological properties, it is this opinion that an attempt to explain all the evidence presented in this investigation using known physiological mechanisms would be improbable. For example, to find an alternative explanation for hair follicles coated with luminescent white material entering the ocular globes would be especially perplexing. The videos showing macroscopic white material being extracted from one area of the plantar surface of the feet and being reabsorbed instantaneously in another location also cannot be explained by known physiological mechanisms in this opinion.

The characteristics of the microorganism described in this investigation have probably existed for a long time, however, unnoticed. The reasons that these phenomena have probably gone undetected until now are listed below:

1. The organism is paradoxically transient. The fact is that approximately 1500 hours of recorded observations have been condensed into a few hours of video. Most observational sessions (usually lasting 3 to 5 hours each) yielded no usable data.
2. The organism mobilizes at very rapid velocities. The organism enters the dermis often via hair follicles in under 33 milliseconds, literally in the blink of an eye. Many video observations were not determined to be significant until the video sequence was examined frame by frame using the Sony Picture Utility software. Without a frame by frame analysis most of the evidence appears to be dust in the wind or hair being brushed off the dermis.
3. The organism is for the most part asymptomatic. Apart from transient paradoxical paroxysmal pruritis and conjunctivitis, the organism is associated with no symptoms. Affected individuals are most likely diagnosed with sensitive skin, psychotic parasitosis and eye strain.
4. The organism is essentially invisible to the casual observer. Indeed, the first 500 hours of video observations yielded only minutes of usable data. It was not until a Welch-Allyn headlamp was recruited as a source of illumination by shear serendipity that the organism was visible. To hinder detection further, the organism often attaches itself to other materials namely hair follicles but may also utilize cotton, feathers, or other filamentous materials. The so-called subcutaneous "worms" turn out to be epithelium with a large area of keratin infiltrated by a dark filamentous material. Again, these "worms" are not worms at all, rather they are some sort of microorganism keratin matrix within the epidermis and affected individuals are probably diagnosed with psychotic parasitosis.
5. The organism only infects a marginalized and more importantly, a criminalized patient population. I consider the lack of research into this microorganism a result of discrimination and apathy towards the most vulnerable population in the developed world. If the organism infected only millionaires, politicians or scientists I assure you the organism would have been discovered years ago. Currently, persons infected with this microorganism are diagnosed with psychotic delusions and prescribed anti-psychotics. The CDC has begun an investigation of complaints of an unexplained dermopathy in non-cocaine using patients.

With the evidence collected, a model of the nature and life cycle of this organism was created. My conjecture is that a microscopic organism enters the body of persons with cocaine in their systems through their epidermis, sclera and cornea. Suitable hosts are probably detected through chemotaxis and the presence of cocaine stimulates the organism to fly in the direction of the source of the cocaine. The microorganism elongates as it flies perhaps to aid in infiltrating the epidermis of the host. As there may be two forms of the microorganism that enter the epidermis they will be dealt with separately one at a time in the following two sections.

One form of the microorganism that enters the dermis is as a filamentous keratin matrix. These are described earlier as coated hair follicles and white reflective and luminescent particles. As this form of the microorganism contacts the surface of the host, it proceeds to enter the host rapidly possibly with the assistance of keratinized epithelium acting as a mechanical debrider and/or through some form of chemolysis of the host's epidermis, cornea or sclera. This rapid breach of the epidermis, together with the internal propulsion system of the microorganism, is the most likely explanation as to why the microorganism enters the host instantaneously, in under 33 ms.

The other form of the micro organism that enters the epidermis is a particle made up of the dark filamentous material without the keratin. These are the so-called "coca mites" that fly off the pipes while cocaine bicarbonate is being vaporized and imbed themselves in the dermis of the smoker's face (see Figure 9.29) and they may even congregate on ceilings waiting for a host that is exuding cocaine molecules from their skin and hair to pass underneath. Unlike the white particles and infected hair follicles which are completely asymptomatic upon entry into the host, the entry into the epidermis of the dark "coca mite" form of the micro organism is accompanied by a pinprick sensation.

Another important difference between the keratin containing and non-keratin forms of the micro organism is that the keratin-containing form of the micro organism enters the epidermis completely and without residual visible evidence of an entry point. The dark non-keratin containing form of the micro organism is seen for days after it is implanted in the epidermis. Any attempts at removal of the dark form of the micro organism are futile and actually makes the appearance worse as the micro organism seems to respond to attempts at its removal by burrowing even deeper.

The differences between the dark and white forms of the micro organism may add information as to the role of keratin in the life cycle of the micro organism. Host interactions with the white keratinized form of the microorganism is for the most part painless indicating that it probably is an analgesic while entry of the dark form into the host is accompanied by a pinprick sensation. The white form enters the epidermis without leaving any evidence of the entry point while the dark filament leaves a black dot at the point of entry. Therefore, the white keratin matrix is probably involved in both avoiding detection and epidermal repair.

Once inside the dermis, the micro organism commences to incorporate keratin into its structure and may also attach itself to organic and possibly inorganic materials perhaps to escape immunological detection and/or to secure a source of nutrients or mobility. Small dilated blood vessels on the sclera, and probably throughout the rest of the body of the host, feed nutrients to the microorganism.

In the presence of vaporized cocaine outside the host's body, the organism is stimulated to exit the body through the dermis, sclera and cornea often at velocities reaching 20 kph. Once outside the body, the organism propels itself towards the source of the cocaine fumes. If the organism happens to land on a piece of cocaine it forms a non-keratinized concentrated mass of dark filaments and awaits a cocaine stimulus. This is the so-called "coca mite" that flies off cocaine pipes and imbeds itself into the epidermis. Once inside the host, the organism most likely begins to incorporate keratin, cocaine and/or other as yet unknown nutrients into a keratin-micro organism matrix and begins to travel throughout the body staying within the epidermis, awaiting another cocaine stimulus.

The organism may also happen to fly into the dermis where it will merely begin to incorporate keratin and other nutrients into its matrix and presumably reproduce itself. Again, the organism and its matrix travel through the body within the epidermis, awaiting another cocaine stimulus.

It is important to keep in mind that this is a model only. The evidence presented here is by no means meant to represent conclusive scientific proof that the micro organism is precisely as it is described in this chapter. This model, however, does explain all the properties of the organism that the evidence demonstrates.

18

Is There a Cure?

The cure may be simple but could also be complicated. The phenomena described here only occur in the presence of cocaine indicating that this substance plays a crucial role in maintaining the unique characteristics of this organism. Eliminating cocaine from the tissues of the host seems to alleviate the most of the symptoms immediately. However, after about five or six days of not consuming cocaine the subject began to develop erythematous patches of pruritis.

Over the next week, the severity of the pruritis escalated to the point that the subject was scratching the affected areas until the epidermis was excoriated. Unlike the paradoxical paroxisms of pruritis experienced in the presence of the vaporized cocaine, the pruritis had a more gradual onset and persisted in the same locations for a few weeks. The most plausible explanation for this is that without the presence of cocaine, the host is able to mount an immunological response against the micro organism and hence the host experiences a persistent progressing pruritis.

Eventually, after a several weeks without cocaine, the pruritis and consequent rash begins to subside indicating a possible reduction of the organism from the system. However, even though the pruritis began to subside, the pruritis persisted, albeit to a lesser degree, in the subject. The ocular symptoms also persist, again to a lesser degree, for an indefinite period. After nine weeks without cocaine, the ocular irritation was greatly reduced but still there nonetheless.

The most plausible explanation for the persistence of the symptomatology after the subject stopped using cocaine is that the microorganism is still inside the subject. Perhaps some proportion of the microorganism is vulnerable to the subject's immune system while another is protected from it. The reason for this is that the microorganism is largely protected from the host's immune system by the keratin within the keratin microorganism matrix. Maybe there is a portion of the microorganism that is not within the keratin matrix and it is this unprotected portion of the microorganism that accounts for the initial persistent localized pruritis.

The initial persistent nature of the pruritis after the cocaine is eliminated from the system indicates some immunologic response to the microorganism.

However, after the initial increase in symptoms, the pruritis and ocular symptoms lessen but persist. Possibly, the elimination of cocaine allows the body to mount an immunological response to the microorganism accounting for the initial permanence of the symptoms. After the microorganisms that are exposed to the host's immune system are eliminated, the pruritis lessens as the amount of exposed microorganism grows thin.

Whether or not the microorganism is eventually completely eliminated from the host is yet to be determined and requires more investigation. The subject of this investigation was suspected of still harboring the microorganism after 9 weeks of abstinence from cocaine. The subject was still experiencing a small amount of symptoms and an examination of his eyes showed a small amount of organism present.

The total elimination of the microorganism from the system may require the elimination of the keratin-microorganism matrix. This could take months, years or never happen at all. The reason the keratin-microorganism matrix may persist indefinitely is that the host's immune system does not eliminate keratin recruited from its own epithelium. Having said that, it appears that the microorganism becomes dormant after the subject stops using cocaine. The symptoms of the sensation that worms were moving through the legs and the scalp tingling no longer occurred.

19

Future Considerations: A Microorganism as an Etiology for Delusional Parasitosis

Unfortunately, spectroscopy and DNA analysis were not attempted due to resource restrictions. These could be performed by others - possibly with financial support from a bio-technology corporation - with some of the samples collected to date.

The future potential uses of this organism are many. As stated earlier, the possible medicinal uses of this organism include antimicrobial medicines and epidermal repair. However, perhaps the propulsion system may lead to other biotechnological uses not yet imagined.

The original goal of this research project was to provide a possible tool in the reduction and hopefully the complete eradication of cocaine usage. It is hoped that researchers more experienced than I, will explore the possibilities accompanying this discovery. At the very least, the exposure of this new infection in cocaine users may discourage first time cocaine users from taking their first blast and may even encourage currently addicted individuals to stop using cocaine. At the most, a means to eradicate cocaine plants may result from DNA manipulation or some other alteration of the micro organism.

Admittedly, the possibility exists that no new microorganism exists. The phenomena presented here could be newly discovered but normal physiological processes. No matter how much evidence is presented, there will be controversy whenever a claim like this is presented. Having stated that, it is imperative that research into these phenomena is given a high priority because the stakes are so high. If even the slightest possibility exists that the evidence presented here is indicative of a new microorganism it must be researched intensely. If it turns out that no organism exists, all that is lost is a little bit of money. On the other hand, if there is a new microorganism that consumes cocaine in mammals then the possibility that it could be utilized to eradicate cocaine from the planet would be infinitely valuable. It would change the world.

Although the original and ultimate goal of this research project was and still is the eradication of cocaine, the research may serve a completely different end. After reading about Morgellon`s disease, I am now fairly convinced that the organism discovered here is the same organism infecting people suffering from Morgellon`s. I think the difference in symptomology may have something to with cocaine being either a catalyst for the microorganism or perhaps it is a vector or both.

To avoid the ethical dilemmas of supplying cocaine to human subjects, further research may involve cats. As stated earlier, cocaine-consuming cats appear to become infected with this microorganism as well. The research could start with feeding cocaine to felines and then infecting them with the microorganism. With a lot of hard work perhaps this new microorganism may herald in a new era with a world free from the ravages of cocaine. At the very least it may properly diagnose people with Morgellon's when presenting with an unexplained dermopathy.

For Product Safety Concerns and Information please contact our EU representative GPSR@taylorandfrancis.com
Taylor & Francis Verlag GmbH, Kaufingerstraße 24, 80331 München, Germany

www.ingramcontent.com/pod-product-compliance
Lightning Source LLC
LaVergne TN
LVHW010622300926
849162LV00001B/28

* 9 7 8 1 1 3 8 4 1 6 7 6 5 *